The Genetic Basis of Abnormal Human Behavior

BY THE SAME AUTHOR

(editor) Prosper Lucas's "On the Heredity of Mental Illness" [1850]

Everything You Need to Know to Understand and Live with Depression

Hereditary Madness: The Evolution of Psychiatric Genetic Thought

The Genetic Basis of Abnormal Human Behavior

Charles David Mellon, M.D.

NEW MEXICO

1997

Genetics Heritage Press

First Edition

ISBN 0-9653362-4-7

Library of Congress Catalog Card Number: 96-77828

Printed in the United States of America

Genetics Heritage Press
02 Sunrise Drive
Placitas, New Mexico 87043

Dedicated

to

Linda Forge Mellon
Katherine Renee Mellon
Barbara Rae Rogers

What I most fear is that in thus attempting to formulate some of the difficult problems of present-day interest . . . I may appear to make at times unqualified statements in a dogmatic spirit. I beg to remind the reader and possible critic that the writer holds all conclusions in science relative, and subject to change, for change in science does not mean so much that what has gone before is wrong as the discovery of a better strategic position than the one last held.

Thomas Hunt Morgan - 1913

CONTENTS

PREFACE 13

1. The Range of Abnormal Behavior in Man 15

DEFINING NORMAL BEHAVIOR AND ABNORMAL BEHAVIOR 15
- What is Behavior ? 15
- What are the Constraints on Behavior 17
- Factors that Determine Individual Human Behavior 18

WHAT IS ABNORMAL HUMAN BEHAVIOR AND WHAT CAUSES IT ? 19

BRIEF DESCRIPTIONS OF HUMAN BROAD BEHAVIORAL PHENOTYPES 22
- Schizophrenia 23
- Bipolar Disorder 25
- Major Depression 27
- Mental Retardation 29
- Dementia 30
- Attention Deficit Hyperactivity Disorder 31
- Anxiety Disorders 32
- Learning Disabilities 32
- Personality Disorders 33
- Substance Abuse 33
- Implications of the Limited Number of Behavioral Phenotypes 34

2. Basic Principles of Genetics and their Implications for the Genetics of Behavior 35

BASIC GENETIC CONCEPTS 35
- The Cell Cycle 36
- The Chromosome Number in Man 37
- DNA: The Molecule of Life 38

The Genetic Code 38
Mutation 40
Chromosomal Organization and Gene Structure 41
GENETIC FACTORS IN HUMAN ILLNESS: MENDELIAN PATTERNS OF INHERITANCE 42
Dominant Inheritance 42
Recessive Inheritance 43
X-Linked Inheritance 44
GENETIC FACTORS IN HUMAN ILLNESS: NON-MENDELIAN PATTERNS OF INHERITANCE 45
The Concept of Predisposition to Disease 45
Simple versus Complex Diseases 45
How to Explain the Genetic Basis of Low Recurrence Risk Diseases 46
How to Explain Familiality in Very Rare Disorders 48
SPECIAL TOPICS IN GENETICS WITH RELEVANCE TO BEHAVIOR 48
Why is there Sex ? 48
Polymorphism and Biological Individuality 49
Anticipation 50
Hardy-Weinberg Equilibrium 51
The Question of Inheritance of Acquired Traits 52
Assortative Mating 52
Parent of Origin Effects 53

3. Behavioral Consequences of Chromosome Abnormalities 55

NOMENCLATURE 55
BEHAVIORAL CONSEQUENCES OF THE LOSS OF A CHROMOSOME 56
Turner Syndrome 56
BEHAVIORAL CONSEQUENCES OF THE ADDITION OF A CHROMOSOME 57
Trisomy 8 Mosaicism 58
Trisomy 13 58
Edward Syndrome 58
Trisomy 21 58
Klinefelter Syndrome 59
XXY 60
XXX 61
BEHAVIORAL CONSEQUENCES OF DELETION SYNDROMES 61
BEHAVIORAL CONSEQUENCES OF DUPLICATION SYNDROMES 66

Implications of Chromosomal Abnormalities on Behavior 66

4. The Behavioral Consequences of Specific Single Gene Disorders 69

Huntington Disease 69
Acute Intermittent Porphyria 70
Phenylketonuria 70
Wilson Disease 71
Tay-Sachs Disease 72
Neurofibromatosis Type 1 72
Myotonic Muscular Dystrophy 72
Metachromatic Leukodystrophy 73
X-Linked Mental Retardation Syndromes 74
X-Linked Hydrocephalus 77
Mucopolysaccharidosis 77
Androgen Insensitivity 78
Single Gene Disorders with Behavioral Implications that are not yet Localized 78
Implications of the Behavioral Consequences of Single Gene Disorders 81

5. Approaches to Studying the Genetics of Broad Behavioral Phenotypes 83

The Role of Family Studies in Studying Broad Behavioral Phenotypes 83
Requirements for Valid Family Studies 84
Risk Assessment 85
The Limitations and Implications of the Family Study Approach 86
The Role of Twin Studies in Studying Broad Behavioral Phenotypes 87
Design and Interpretation of Twin Studies 87
Misconceptions and Limitations of Twin Studies 89
Studies of the Offspring of Discordant Twins 92
Using Twins to Study Comorbidity 92
Adoption Studies 92
Design and Interpretation of Adoption Studies 93
Limitations of Adoption Studies 93

High Risk Studies 94
Design and Interpretation of High Risk Studies 94
Segregation Analysis 95

6. The Genetics of Broad Behavioral Phenotypes 97

The Genetics of Mental Retardation 97
Epidemiology of Mental Retardation 97
The Causes of Mental Retardation 97
Implications for the Genetics of Abnormal Human Behavior 101
The Genetics of Schizophrenia 103
Family Studies 103
Implications of Family Studies in Schizophrenia 105
Twin Studies in Schizophrenia 107
Monozygotic Schizophrenic Twins Raised Apart 110
Use of Twin Studies to Establish Diagnostic Criteria 111
Schizophrenia Rate in Offspring of Discordant Monozygotic Twins 111
Conclusions from Schizophrenia Twin Studies 112
Adoption Studies in Schizophrenia 112
Attempts to Define a Mode of Transmission in Schizophrenia 114
Conclusion 115
The Genetics of Bipolar Disorder 116
Family Studies 116
Twin Studies in Bipolar Disorder 117
Mode of Transmission 119
Unipolar Disorder (Major Depression) 120
Primary Degenerative Dementia of the Alzheimer Type 121
Alcoholism 123
Twin Studies in Alcoholism 123
Adoption Studies of Alcoholism 123
Stuttering 124
Family Studies of Stuttering 125
Twin Studies in Stuttering 127
Mode of Transmission of Stuttering 127
Tourette Syndrome 129
Pervasive Developmental Disorder (Autism) 131
Attention Deficit Hyperactivity Disorder 135
Specific Developmental Disorders (Learning Disabilities) 136

THE GENETICS OF PERSONALITY DISORDERS 137
CONCLUSIONS FROM THE STUDY OF THE GENETICS OF BROAD BEHAVIORAL PHENOTYPES 139

7. Strategies to Localize the Genetic Factors in Broad Behavioral Phenotypes 141

THE ROLE OF ASSOCIATION STUDIES 141
LOCALIZING GENES WITH CLINICAL HINTS FROM UNIQUE CASES 143
The Role of Unique Chromosomal Abnormalities 143
Case Reports or Disease Associations 144
THE ROLE OF LINKAGE STUDIES IN LOCALIZING GENES 144
The Revolution in Gene Mapping Using Polymorphic DNA Markers 145
Linkage Basics 145
Experimental Procedure of DNA Marker Based Linkage Studies 148
Statistical Analysis of Linkage Data 149
Strategies for Linkage Studies 150
The Consequences of Establishing Linkage 151
Potential Limitations of Linkage Studies 153
The Future of Linkage Studies in Psychiatric Genetics 153

8. The Current Status of Attempts to Localize Genes for the Broad Behavioral Phenotypes 155

THE SEARCH TO LOCALIZE A GENE FOR SCHIZOPHRENIA 155
Association Studies 155
Unique Clinical Cases with Hints About Localization 155
Candidate Genes 156
Linkage Studies in Schizophrenia 157
Conclusion 159
CURRENT STATUS OF LOCALIZING GENES IN BIPOLAR DISORDER 160
Association Studies 160
Linkage Studies in Bipolar Disorder 160
Conclusion 171
OTHER ATTEMPTS TO LOCALIZE PSYCHGENES 172
Current Status of Attempts to Localize Genes for Alzheimer Disease 172
Current Status of Attempts to Localize Genes for Autism 173

Current Status of Attempts to Localize Genes for Alcoholism 173
Current Status of Attempts to Localize Genes for ADHD 174
Current Status of Attempts to Localize Genes for Panic Disorder 174
Current Status of Attempts to Localize Genes for Stuttering 174
Current Status of Attempts to Localize Genes for Dyslexia 174
SPECULATIONS ON THE FAILURE TO LOCALIZE GENES FOR BROAD BEHAVIORAL PHENOTYPES 174

9. Special Concepts in Psychiatric Genetics 177

VARIATION IN BEHAVIORAL DISORDERS AND THE SPECIAL ROLE OF DEVELOPMENTAL PLASTICITY 177
Variation 178
Sources of Phenotypic Variation in Behavioral Disorders 179
The Effect of Variation on Final Phenotypic Outcome 183
The Evolutionary Importance of Developmental Plasticity 186
THE CONCEPT OF PHENOCOPIES AND ITS ROLE IN BEHAVIOR 187
Mercury 188
Phenylketonuria 188
Alcohol 188
Lead 188
The Future of Behavioral Teratogenesis Research 189
Is Congenital Deformation a Source of Abnormal Behavior ? 189
The Role of Psychological Life Events in Creating Abnormal Behavioral Phenotypes 190

10. Implications for the Practice Psychiatry 193

IMPLICATIONS FOR IMPROVED DIAGNOSIS AND TREATMENT 193
A Genetic Oriented Family History 194
The Physical Exam: The Psychiatrist as Dysmorphologist 194
Using Genetic Related Lab Tests 195
Treatment: General Principles 195
Treatment: The Psychiatrist as Genetic Counselor 196
Technique of Psychiatric Genetic Counseling 197
Problems in the Counseling of Behavioral Genetic Disorders 198
Appendix A: Behavioral Genetics Service Organizations 201
References 203
Index 253

Preface

This book is an attempt to define the central role that genetics plays in abnormal human behavior. The argument is supported with evidence accumulated by thousands of individuals over the past century. Areas of current controversy are identified and speculated upon. References are as detailed and exhaustive as possible. This book also departs from previous behavioral genetics monographs in that it is focused exclusively on human studies and avoids ethological based research. Furthermore, every effort has been made to use only behavioral genetic examples to illustrate genetic points, and to avoid non-behavioral examples.

The book follows a specific plan. Chapter one demarcates normal and abnormal behavior. Chapter two discusses advances in genetics and the implications of each advance on the study of behavior. Chapter three and four cite specific correlations between various genetic abnormalities and the resulting abnormal behavior. Chapter five discusses the approaches to studying broad behavioral phenotypes while chapter six reveals what is currently known about those phenotypes. Chapter seven and eight outline strategies currently being used to further localize genetic factors in broad behavioral phenotypes and document the findings so far. Chapter nine reflects some special concepts in psychiatric genetics. Finally, chapter ten discusses the potential impact of these findings on diagnosis and treatment in psychiatry and the future of psychiatric genetic research.

My motivations for writing this book are several. I have a deep respect for the field and the answers it provides about mental illness and the philosophical implications that follow from it. I also feel a need as a practicing psychiatrist to help deflate the myths, misconceptions, and other unfortunate legacies that interfere with an unbiased view of the field psychiatric genetics. Finally, I wanted to highlight the accomplishments of the various individuals, often unheralded and misunderstood, who labored over decades exploring the role of heredity in mental illness.

This book was written for a wide audience. Every effort has been made to make the book as readily accessible to the interested layman as it is to the expert in the field. This book is particularly designed for physicians, psychiatrists, psychologists, geneticists, social workers, and genetic counselors. However, families with mentally ill relatives or individuals with mental illness may also find this work useful in understanding the biological basis of the disorders.

As in any undertaking of this magnitude, a few caveats are in order. To those experienced with the psychiatric genetic field, there may be some surprise to see relatively few mathematical equations, complex graphs, or tables in this book. This was done for a specific reason. I feel that most individuals are hampered in their understanding of this important field by the confusing and overwhelming appearance of the technical arguments involved in psychiatric genetic research. In reality, almost all of the concepts involved in psychiatric genetics can be understood and appreciated without delving into the hairsplitting details or complex mathematical analysis.

Writing this book has been a challenge and an education. I would like to thank the people who helped make it possible. First of all, my wife Linda Forge Mellon, Ph.D. for her constant support and helpful editing skills. I would also like to thank my early mentor H. Eldon Sutton, Ph.D. for helping spark my interest in human genetics. The National Alliance for Research on Schizophrenia and Depression (NARSAD) for their fellowship award and the National Institute of Mental Health (NIMH) for their early support.

Chapter 1

The Range of Abnormal Behavior in Man

DEFINING NORMAL BEHAVIOR AND ABNORMAL BEHAVIOR

What is Behavior?

In the broadest sense, behavior can be defined as "any or all of an organism's total activity, especially that which can be externally observed." In this definition, behavior can range from simple motion to complex thoughts. All animals from amoebas to man exhibit behavior. Regardless of the organism, however, the same basic relationship exists between the organism and its environment. The organism exists within a finite environment and must interact within that environment to meet its needs. This critical synergism between the organism and the environment depends upon a net derivation of energy by the organism. If the organism derives a net energy from the environment, the result is growth. If a balance is maintained, there is homeostasis. Finally, if the organism loses more energy to the environment than it gains, the result is death. Life, in this regard, can be seen as a continuous fight against a loss of energy or entropy. Behavior, therefore, is best seen as mechanisms that serve to increase the organism's ability to derive net energy from the environment and thus to maximize its survivability. In a unicellular organism, a behavior that can increase survivability can be as simple as motion. The ability to actively seek out food sources or to evade noxious stimuli has obvious advantages over a passive non-mobility. In organisms like man, behavior has evolved to the point that even such complex behavior as language and symbolic reasoning have become essential to survival.

When an organism's behavior allows it to survive and thus reproduce at a greater rate than other organisms then that behavior gives it a selective advantage. This process pushes behavior toward an ever more efficient means of maximizing an organism's survival. All normal behavior is purposeful. It has evolved to increase the survival of the organism. All behavior, no matter how simple or complex, is an attempt by the organism to control and manipulate its environment. However, since behavior evolves to match the environment that the organism's ancestors lived in, even the fittest behavior may not lead to survival if the environment exceeds the organism's range of adaptive capacity. The differential rate of change between biological adaptation and environmental change may lead to the persistence of anachronistic behavior. No matter how purposeful behavior is, if it is not matched to its environment, then it may not lead to survival. This principle implies that novel situations can overwhelm the organism's capacity to use adaptive behavior. In humans, this might involve sleep deprivation or extreme stressors. Under these conditions, the individual's actions may not serve to increase its survivability. An example is the person who is faced with overwhelming problems and commits suicide. It is very hard to argue that suicide increases an individual's survival.

Behavior increases survival by increasing the efficiency by which an organism can procure what it needs from the environment. There is, however, a downside or counterbalance to the adaptive role of behavior. Since all behavior requires an input of energy, and require complex sensory and cognitive apparatus, *all behavioral systems are subject to a vast number of potential errors that can seriously disrupt the organism's ability to adapt*. Examples of possible sources of damage or error in behavioral mechanisms are basically the list of ways that the brain can be injured or malformed. For instance, congenital or genetic malformation of the brain will seriously impact the individual's ability to use a full range of adaptive behavior. Conversely, even a well formed Central Nervous System(CNS) cannot produce normal behavior if damaged by trauma or infection or if influenced by psychoactive substances.

Behavior in man is mediated through a finite number of observable avenues; *motion* – in the form of the movement of the muscle groups in smooth, purposeful activities, *speech* – as the end product of complex interplay between the motor, speech, and higher cortical centers, and *activity* – as the complex enduring series of decisions and social interactions that comprise the personality. Though few in number, these avenues can combine to form a wide range of behavior. Yet, behavior, even in man, is not endless in range and it is constrained by many factors.

What are the Constraints on Behavior?

Since behavior is an epiphenomenon of the organism, all behavior must be a product of the same chemical and physiological laws that govern all matter. No matter how complex behavior becomes, it ultimately must follow the constraints of the first and second laws of thermodynamics. This is a very important concept in understanding behavior. The idea that behavior cannot be reduced to the laws of physics is best summarized by the concept of "vitalism." Vitalism is the ancient belief that living organisms are somehow exempt from, and do not obey the laws of physics. The rise of science since the Renaissance has repeatedly pushed back the concept of vitalism. Anatomy revealed an ordered and understandable complex of organs and systems. Biochemistry revealed logical sequences of chemical reactions that could be duplicated *in vitro*. Genetics revealed that the code of life itself follows the laws of physics. There is every reason to believe that behavior with all of its complexity also follows logical rules within the laws of physics. Because of its vast complexity, however, behavior remains one the last bastions of vitalism in science. A second constraint on understanding behavior is the evolutionary sequence within which it has come to be. No concept is more important for an understanding of behavior of any type than that of evolution. Evolution and natural selection demand that behavior be purposeful (increase survivability), and conversely that unpurposeful or maladaptive behavior be selected against.

If behavior follows the laws of physics and is a product of evolution, a logical question to ask is what physical or anatomical structures in the organism control behavior? In simple organisms, a neural network with simple reflex arcs is responsible. In higher animals, behavior is much more complex and has required ever more sophisticated levels of organization. The result in higher animals, including man, is the Central Nervous System (CNS). It is through the CNS that all thought, fears, and ideas are developed. The simple observation that all behavior is a result of the chemical and electrophysiological activity of the brain can be difficult to keep in focus, but no meaningful, scientific understanding of normal or abnormal human behavior can be obtained unless this basic concept is understood.

Factors that Determine Individual Human Behavior

Several factors determine the behavior of the individual. The first factor is the genetic endowment of the individual. Each individual begins the process of development at conception. At the point of uniting of sperm and egg all of the genetic information necessary to develop the organism is present. The 46 chromosomes, which contain the 50,000 - 100,000 genes, have the complete plan for a unique individual that will never be repeated. The information present in the code is the product of countless numbers of generations, each surviving to reproduce. This process ensures that the vast majority of individuals will have a genetic code that is primed to develop an organism optimized for survival. However, the process is not failproof, and because of its tremendous complexity, errors commonly occur.

The second factor is prenatal development. Starting with a single cell, the human organism undergoes enormous growth. It is somewhat counterintuitive, but more events of importance to the health and behavior of an individual occur during the nine months of prenatal development than occur any time thereafter. This massive period of growth and embryogenesis is a time when all organs and systems are developed. The brain undergoes a phenomenal growth during this time. Anything this complex has a high likelihood of errors. A combination of errors in the genetic code and development is apparent in the fact the 40% or so of all conceptions are spontaneously aborted due to non-viability. It is reasonable to assume that the complex developing CNS is subject to many errors, both subtle and catastrophic.

The third factor is "chaos." Simply stated, the brain contains trillions of connections and cells. It is not reasonable to believe that these connections are closely regulated in development. This allows for a tremendous amount of randomness in development. This randomness is nowhere more apparent than in identical twins who have the same genetic code, but are often diverse in their behavior. Development does not stop at birth. The individual goes from the protected parasitic environment of the womb into a much more complex and unprotected separate existence. It is a time of great demand on the individual's ability to adapt. This is reflected in the fact that the infant death rate is the highest death rate until the individual reaches old age. Malnutrition, infection, and trauma all can interrupt the orderly development of the growing organism, leading to damage or abnormal development of the brain, thus leading to abnormal behavior.

The fourth factor is environment. From the moment of conception the organism exists within an environment. The environment can be defined as all the external factors that influence the development of the organism. For this reason it can be stated that environment is the playing field upon which the individual's genetic potential is played out. Consistent environments that have been traditionally present to the organism throughout recent evolutionary history will be the optimal match for the organism. Extreme environments may exceed the built in capacity of the individual to adapt as determined by the genetic code and previous development. This will lead either to the death of the organism or in lesser amounts to disease. It is important to realize, however, that a normally developed human being has a tremendous adaptive capacity. Small environmental insults that lead to large adaptive problems indicate an abnormal baseline range of adaptive capacity in the individual. In other words, the more impaired a person's biological capacity for adaptation, the more susceptible they will be to even minor environmental insults.

What is Abnormal Human Behavior and What Causes It ?

Why is there abnormal behavior? What factors are involved in its causation? Humankind has attempted to answer these questions throughout recorded history. Theory upon theory have at times seemed only to make the answer even more distant and unobtainable. Perhaps because of this uncertainty, the issue has been clouded behind obscurant and vitalistic interpretations that add little to our knowledge and serve only to confuse and dilute the question. Surely, despite its complexity, the veil blocking our understanding of abnormal behavior is falling just as it has for so many other seemingly unanswerable questions in science.

All behavior exists for a reason. Very little, if any, behavior in normal animals is wasted. Even "play" behavior in mammals serves as modeling and testing behavior for more critical situations. As important, if not more important, as any physical adaptation, behavior in all its complexity serves the penultimate goal of survival of the organism so that the organism can reproduce. How then do we explain abnormal behavior? Clearly, this behavior does not increase the survivability of an organism. Why does it exist? It must be understood from an evolutionary perspective. Abnormal behavior in an evolutionary sense is essentially a failure of adaptation. However, this failure of adaptation is grounded in the genetic and physiochemical development of the individual.

If evolution has driven behavior in man, it has done so through the formation of ever more sophisticated brain structures. All human behavior is dependent upon a normally developed active, functioning brain. We know that destruction of the brain has behavioral consequences. Only the most vitalistic person would deny the brain its role as the keeper of man's intellect and personality. Disturbances of normal brain development and function have direct effects on behavior. If the developing brain, for instance, is deprived of oxygen beyond the critical few minutes there are clear effects on motor and cognitive skills. Likewise, toxins, infections, and high temperature can effect behavior. The brain mediates behavior, and the brain is by far the most complex of all the human organs and also the most rapidly evolving organ. In the past three million years of human evolution, amazing changes have occurred to our brains. The volume has literally tripled with over two kilograms of new brain cells added. These changes allowed for the expansion of the association cortex with a resultant capacity for abstract reasoning and language. This, in turn, has liberated man like no other animal from the rule of instinct and innate patterns of reasoning. Modern man has tremendous behavioral flexibility, which is ultimately a product of his genes, his language, his culture, and his education. However, in every language, in every culture, in every level of education there appear individuals whose behavior is abnormal.

The answer to the question of why there is abnormal behavior is straightforward. Behavior is the product of a series of events that begin with the genetic information at conception. This information is processed through development and ultimately tested by the environment. Opportunities for error abound. So many, in fact, that it does not seem possible that the process can be completed without error. These errors interfere with the final functioning CNS thus producing abnormal behavior.

This is not to say that environment plays no role in the development of abnormal behavior. Genetic and developmental factors define the physiologic limitations of the individual's behavioral repertoire. Learning can determine the actual expression of behavior within the physiologic range. In lower animals, behavior is determined by innate instinctual drives that sacrifice adaptability for precision. In general, this serves the evolutionary needs of these organisms. However, in higher animals, especially primates, the length of time devoted to maturation allows for a variety of environmental inputs. From an evolutionary perspective this allows for even greater adaptability of response to changing environments. From a physical perspective, man is weak and unprotected compared to other animals. His adaptive advantage lies in his ability to manipulate his environment. The price for this adaptability is a prolonged period of maturation, not only in utero, but also in social development. Even among the primates, the delay in attaining adult reproductive status is especially prolonged in humans. At birth the human child has less than one fourth of the total brain mass that he will have in adulthood. However, the newborn has all the neurons that she will ever have. This long period of maturation allows for a vast number of cultural influences, which in turn influence behavior. The advantage of this strategy is that man can learn enough to exist in many different environments. The disadvantage is that the knowledge gained during this long period may or may not be consistent with that needed for adaptation in adult life.

If behavior is critical to survival and reproduction, how can a species survive if abnormal behavior exists? The answer is that reproductive rates are far higher than needed just to replace members that die. This allows for "wastage" without the risk of eliminating the species. Only under the most radical of environmental changes will the species itself be in danger. Almost by definition, error rates within a species cannot effect more than a small portion of the overall population.

How does abnormal behavior manifest in man? The answer to this question is that abnormal behavior in man seems to fall into distinct patterns. These patterns of abnormal behavior can be best described as "broad behavioral phenotypes." Broad signifies that large numbers of individuals end up being described adequately by one central cluster of abnormal symptoms. Phenotype is a genetic term meaning how the genetic influences are expressed in the individual. Broad behavioral phenotypes then are descriptive representations of abnormal behavior that encompass large segments of the individuals with abnormal behavior. There are only a limited number of such broad behavioral phenotypes in abnormal human behavior.

BRIEF DESCRIPTIONS OF HUMAN BROAD BEHAVIORAL PHENOTYPES

Schizophrenia

Schizophrenia is a term that refers to a broad behavioral phenotype characterized by a pronounced abnormality in the thought processes of those afflicted. Affected individuals have a fairly distinctive pattern of long term behavior. The pattern of an individual's behavior over a many year time period is referred to as the "longitudinal course" and is of paramount importance in schizophrenia. Typically these individuals have trouble with social relations and social expectations that are usually first noticed in adolescence as demands for adaptive behavior increase. Isolation and withdrawal often proceed the more severe symptoms of auditory hallucinations, delusions (false beliefs), and paranoia. Early onset of this disorder is the rule, although a person may escape formal diagnosis for many years. Kraepelin felt that the early onset was of central importance as reflected in his term "dementia praecox" (or premature dementia). Typically individuals with schizophrenia have a waxing and waning but downhill course. Rarely will they return to the level of function they were at prior to their most recent exacerbation. The longitudinal course, therefore, reveals early problems with social withdrawal, periodic psychosis, and generally poor adaptive function in society. Often family members and others will step in and provide assistance allowing for the person to maintain a reasonable lifestyle. However, sometimes due to lack of support or severity of the condition, many individuals become marginalized and live difficult, brutal lives such as those of the "street people" found in most large urban areas. Current estimates place the prevalence of schizophrenia at around one percent or slightly less. The incidence seems to be fairly constant throughout the world. Treatment is based on the use of neuroleptic agents, which block dopamine receptors, but it is generally agreed that the treatment is only moderately effective, mostly controlling hallucinations and delusions while doing little for the social withdrawal and adaptive disability. The cost of schizophrenia is incalculable.

Individuals with schizophrenia exhibit a great deal of variation in both severity and symptoms. Mild forms of the disorder are compatible with employment, marriage, and child raising. More severe forms lead to incapacitation and occasionally to institutionalization. The following is an example of schizophrenia:

> As a child, John seemed a little more sensitive than his sister, but he did well in school and played basketball well enough to be in the first string throughout junior high school. He had a couple of good friends and liked to collect comic books. His parents noticed that about the onset of his sophomore year in high school he began to spend more time in his room and rarely had any friends over. His personal hygiene began to deteriorate to the point where he constantly had to be told to shower and wear clean clothes. His teachers reported to his parents that his grades were dropping except in art class where his teacher was very impressed with his introspective form of painting. His parents became concerned and brought John in for psychiatric evaluation wondering if he was depressed. Mild depression was noted and John was placed on antidepressants and given individual counseling. John's behavior improved slightly, but over the next two years John continued to isolate

himself. He stopped playing basketball, and spent more time on his comic book collection. Sometimes his mom would hear him talking in his room when no one else was there. His parents were worried that John might be using drugs, so they again brought him in for evaluation. John admitted trying marijuana a couple of times, but denied other drug use. He did confide to the psychiatrist that he was hearing the voices of some of his comic book heroes. They would tell him to do things or comment on what he was doing. He knew it was not normal to hear voices so he had not told anyone. Medications were discussed with John's parents, but they preferred to continue antidepressants and counseling. John ultimately graduated from high school with marginal grades, but received a high enough SAT score to qualify for a junior college at the other end of the state. His parents thought it would be good for John to be on his own and "grow up a little." John moved to the college town and his parents rarely heard from him. On a visit, they found him isolated in his dorm room, not attending classes, with unkempt hair and clothes. He asked to come home. After six months at home, mostly staying in his room, his parents pressured him to take a job at his dad's company. After a few days, John asked to quit, but his parents insisted that he continue. One day shortly after, John disappeared. He took the family car and a little money. His parents heard nothing from him for one month until they received a call from a psychiatric hospital in another state. John had been hospitalized after he was found wandering around talking to himself, stating that he was a character in one of his comics and that the "forces of evil" were after him. He admitted that he destroyed the car radio because "they" were using it to read his mind. After three weeks in the hospital, John was discharged to his parents' home, where he spent the next several years living until he was placed in transitional living. He qualified for social security disability. He continued to occasionally read comic books. On a couple of occasions he stopped his medications and was briefly rehospitalized. As his parents got older, his sister began to look after him.

Schizophrenia appears to be a disease of cortical function. It has long been noted that there is a disturbance in the ability to determine relevant from irrelevant sensory input in these individuals. Stimuli of any type is hard for them to process, so they tend to withdraw from arousal. Avoiding stress and other stimuli is a compensation that lessens symptoms. This is not universally understood, and there are still many theories that emphasize the isolation and withdrawal as being involved in the causation of schizophrenia. Psychotherapy has proven to be of little more than supportive value. Most treatment programs now emphasize case management to help individuals with activities of daily living. Psychosocial rehabilitation programs that prevent extreme isolation and increase social connections are also helpful. Regardless, schizophrenia, in all but the most mild cases, continues to be a profoundly destructive disease that is poorly understood by the general population.

Bipolar Disorder

Individuals with this phenotype have a prevalent disorder in the regulation of mood. Characteristic of this disorder is the presence of a least one episode of mania. Mania is a state of heightened mood in which a person begins to lose perspective on the normal social and behavioral guidelines that control their behavior. Judgment becomes impaired leading to impulsive behavior, such as excessive spending, promiscuity, and a sense of omnipotence. The condition is periodic. Long intervals of normal function are common in between episodes of decreased function. The incidence is about one percent of the population. The condition is highly variable in severity and duration with a number of distinct subtypes described. Mood dysregulation is the hallmark. Treatment is with mood stabilizing medications, such as lithium or tegretol. Almost all cases can be stabilized medically, but the illness predisposes to poor judgement, which often leads to the individual stopping medications and relapsing. The following is a clinical example of bipolar disorder:

> [A family member's description of his mother's behavior.]
> I would like to have the chance to explain my mother's

condition more fully. What I have said does not differentiate my mother from an eccentric, but otherwise normal person who is perhaps a little over concerned about the environment. However, her concern for toxins is by no means normal and has become an all encompassing obsession, which has made it impossible to function normally as a responsible human being. My mom believes that she has been chosen to initiate a war against lead and other toxins. She recognizes that everyone does not see things her way, and she wants me to tell you my side of the story even though she realizes our stories will differ. My mother has perhaps never enjoyed full mental health. She has had bouts of depression at numerous points in her life. However, her present condition is quite recent and came on rather suddenly. Beginning almost two years ago, my mother began having dizziness and lack of energy. She went to a doctor and had a normal MRI. Shortly after this, my mother met a man that was strongly opposed to the use of lead in household items. She has since become obsessed with lead toxicity. My mom used to be a teacher and had some background in science. She has been reading vast amounts of medical and new age literature and has synthesized them into bizarre theories with lead as the center of the battle between good and evil. My mother used to be a very rational person, but she has become increasingly irrational. When she reads things, especially medical literature, she seems to selectively interpret it into her belief system. She believes that doctors have been "brainwashed" into a conspiracy to deny the effects of lead on peoples health. However, if she finds even a clause that might provide evidence for to her theories she will point to that clause as foundational evidence for her claims while ignoring the majority of the more damaging evidence. She has become dogmatic and dangerous in her beliefs. For instance, she became convinced that hydrogen peroxide destroyed lead and began drinking concentrated solutions of hydrogen peroxide. She only stopped this when she read somewhere that fresh fruit contains hydrogen peroxide and stated to eat only fresh fruit. Her concern for lead toxicity has become the focus of my mom's life. She talks about very little except it. She has lost most of her friends because of her prolonged lecturing.

> She had to quit her administrative job of 25 years and lost all of her seniority. My mom decided that she needed to "purify her environment." She felt she could tell if something was toxic by holding it for a few seconds. She systematically went through the house and threw away nearly everything including things that were not hers. She would not even let us sell them or donate them, she took them straight to the dump. Previous to this time my mother was always careful with money, but this has changed. She pulled out all of her $ 40,000.00 retirement and to our horror spent it in three months. She bought me a car that I didn't need saying that it would be easy to convert to solar power when the time came. My mother spent $ 4,000.00 on a credit card with no means of paying it off. She doesn't seem to be bothered by any of this saying that when she won her case against the corporations using lead she would be rich. She is always starting projects and abandoning them. Some of these projects involved making money, but she quits before she starts. She is still self sufficient at a basic level. Her grooming and eating habits seem to be all right, but she is more and more reluctant to discuss what she is doing to herself and gets defensive if we ask.

The fact that bipolar disorder responds to lithium and some antiseizure medications strongly suggests that these medications simply slow down overall cortical functioning leading to slower cognition and limitation of mood dysregulation. Psychotherapy is of little help in bipolar disorder, but compliance can be greatly improved with the establishment of a strong therapeutic alliance.

Major Depression

Individuals with a depressive phenotype are usually prone to recurrent bouts of depressive episodes. Such individuals periodically have states of decreased mental energy and enthusiasm. Neurovegetative symptoms are common such as weight loss, insomnia, lack of energy, and difficulty concentrating. Cognitive impairment includes feelings of worthlessness, hopelessness, loss of pleasure, and suicidal thoughts. Variation in severity is common with some individuals experiencing debilitating episodes of depression where they cannot work or function in a social setting, and others having brief periods of down feelings with little interruption in their normal routines. The following is a clinical example of an individual with recurrent mild to moderate depression:

> Sarah was described as a moody child by her parents. By the time she was in high school she wondered what the point was to life. She became very interested in philosophy and psychology and experimented with several religions. She found the early years of college to be liberating and excelled at school although she was seen as a "bookworm" by her friends. About the time of her senior year in college she felt very tired all of the time and for the first time got a few "C" grades. A friend suggested that she get some counseling. She spent about a year in counseling, which she felt helped her understand herself much better. After college she applied for and was accepted to medical school. Medical school was quite demanding on Sarah, but she adjusted fairly well. However, at the beginning of her third year of medical school she began to feel tired and unmotivated. Her sleep worsened almost every night. Some nights she would get only 3-4 hours of sleep. She lost 15 lbs. Late at night she wondered if it was "all worth it," and sometimes imagined what it would be like to be dead. One day a clinical supervisor pulled her aside and asked her why her performance was so uninspired given that she was one of the top students in her class. She began to cry and told the supervisor about her concerns that she was a "fraud" and really not capable of being a doctor. She was concerned that now that she was seeing patients, even as a student, her incompetence would hurt someone. The supervisor knew that she was a talented, intelligent student. A referral was made to the student health service for a psychiatric evaluation. A full medical checkup was done, but it revealed no problems. She was placed on a trial of antidepressants and within a few weeks she began to feel that she was making progress with her clinical skills. She graduated in the top 10 percent of her class and was accepted at a prestigious residency. She stopped the medication on her doctor's advice. Two years later she began

> to experience the same symptoms, but this time she recognized them as depression and restarted antidepressants.

It is very difficult to establish the prevalence of depression. One of the main reasons is that brief situational depressive episodes are often counted as depression. These brief episodes are part of the normal human adaptational response. They bear only a superficial resemblance to true major depression. There is a clear difference in abnormal human behavior as found in recurrent major depression and human unhappiness as is found in brief situational depressive episodes. The former comes on for no reason and is resolved only with time or treatment. The latter is directly related to some loss and can usually be resolved by simple changing the current situation. This point is critical. No logical discussion of the role of genetics in depression can proceed without an understanding of this point.

Mental Retardation

This broad behavioral phenotype is the best studied and best understood of the broad behavioral phenotypes. About 1.2 percent of people are mentally retarded. A few are severely retarded, some are moderately retarded, but the vast majority are mildly retarded. As a group, they exhibit global delay from early childhood. Normal childhood milestones are delayed. Often the problem becomes obvious as schooling starts. Typically these individuals have minimal reading and language skills. Social awareness is lacking sometimes worsening attempts at social integration. The following is a clinical example of mild mental retardation:

> Oscar was born underweight and six weeks premature. He was difficult to arouse and was kept in the hospital several days before being sent home. Breast feeding was very difficult because of poor suck response. By 18 months, Oscar was not yet walking, but he was crawling. His first word came at 20 months. At school Oscar had trouble understanding, so he was placed in a special education class where he did much better. After high school he took a job as a mail clerk where his job was to deliver mail to the various offices and do other odd jobs. He lived at home and walked to work. After ten years of

> working at the same job, his mother died. He was unable to manage the household chores and was placed in a nearby supervised living home, but he still goes to work.

Numerous specific causes of mental retardation are now known, however, from a clinical point of view the observed behavior is very similar despite the varied causes. Some causes of mental retardation are preventable, but virtually none are reversible once the symptoms are well in place.

Dementia

This broad behavioral phenotype is best described as the global loss of cognitive function that previously existed. It is usually a slow degenerative process over a number of years. Clinical signs include forgetfulness, difficulty with problem solving, loss of spatial ability, declining personal care, irritability, and suspiciousness. Some forms of dementia occur very early in adult life, some occur in middle age, but the majority occur late in life. The following is a clinical example of a somewhat typical case of dementia:

> Harriet is 68 years old. For 40 years she taught high school English until she retired at age 60 to travel with her husband Arnold. For the past several years, Harriet has had trouble with forgetting things. At first she would misplace small items or forget a phone number, but as the years progressed she began to forget that she was cooking and even the names of her grandchildren. Her husband reports that she stopped driving three years ago after getting into a minor accident. For years she had balanced the family checkbook, but after several bounced checks, she gave up the checking account to her husband. In the evening, Harriet sometimes gets worried about Arnold if he is not at home. On one recent occasion she called the police to find him. Harriet has played the piano for many years and she still plays the old songs very well, but has not learned a new one in years. A couple weeks ago the family got a scare when Harriet disappeared from home. They found her half a mile away where she appeared scared and tired.

> Apparently she got confused and thought she was supposed to go to her mother's house, although her mother had been dead for years. The family has decided that when Arnold is not at home, their daughter Anne will come over and watch Harriet.

Some cases of dementia are reversible if due to specific metabolic or neurologic causes. Dementia of the Alzheimer type, which represents the vast majority of dementia, remains untreatable.

Attention Deficit Hyperactivity Disorder

This broad behavioral phenotype of children involves excessive mental and motoric activity. It is manifested as fidgetiness, difficulty paying attention(especially in school), inability to complete assigned tasks, lack of awareness of danger, and verbosity. The following is a clinical example:

> Julian's mother swears that Julian kicked more when she was carrying him than any of her other children. By age one she knew that Julian was different. He had been walking for three months and would chase the other children around. By school age, his teachers reported that Julian was bright, but he would never sit still and he often seemed to ignore instructions. By third grade, the school asked for Julian to be evaluated. His mom reported that they had similar problems with Julian at home. He never seemed to listen and he never sat still even to watch TV. He seemed to be fearless, often doing any dangerous thing that other children would dare him to do. He talked excessively and often interrupted others. Testing showed intelligence in the high average range. He showed no symptoms of a mood disorder. He was placed on ritalin and his school performance improved dramatically as did his behavior at home.

As more research is being done on this disorder, especially long term follow up studies, it appears that these symptoms continue into adulthood in a slightly modified form. Treatment with stimulants has been very helpful. The theory is that these individuals actually have low cortical arousal and do better with extra stimulation.

Anxiety Disorders

This broad behavioral phenotype describes a range of problems involving a pervasive sense of uneasiness and a feeling that "something bad is going to happen." They include two major subtypes. *Generalized anxiety disorder* is a syndrome involving constant unnecessary worrying and feelings of uneasy. *Panic disorder* is a syndrome where individuals have brief episodes of intense panic where their pulse quickens, their palms sweat, they may have trouble breathing, and the feel like something terrible is happening. The following is a clinical example of panic disorder:

> Kelly is a 35-year old secretary. She describes herself as always having been a nervous person, but about a year ago she had an episode where she felt like she was dying. Her heart raced, she became sweaty, she felt like she was passing out. It lasted only a couple of minutes. She rushed to the emergency room fearing she was having a heart attack. A complete physical revealed no abnormalities. The episodes recurred several times and she was finally referred to a psychiatrist. She admitted to the doctor that she quit her job because of the episodes and now she rarely leaves her house. When she does leave she does not want to drive and somebody she trusts always has to go with her. She was placed on imipramine and has had virtually no episodes in the past two years. She has returned to work with some difficulty.

Learning Disabilities

This broad behavioral phenotype contains a number of described syndromes involving delay in specific areas of functioning such as reading, math, expressive language and receptive language. The symptoms start in childhood, but often persist into adulthood. The most recognized example is dyslexia, where otherwise intelligent individuals have a disproportionate problem with reading. These disorders are not due to obvious problems like decreased vision or inability to hear. The implication is that there is a problem with integrated CNS functioning in relation to the specific task. The diagnosis is often made by identifying the bottom three to five percent on standard tests. An approach that is considered very questionable and likely includes many individuals without a specific problem, but who just happen to be at the bottom of the bell curve. A better approach would be to identify individuals only by clinical complaint and specific functional testing. Therapies have been developed that involve specialized training in recognizing and treating altered patterns of learning.

Personality Disorders

Personality disorders are a group of conditions where an individual has persistent problems with interaction and functioning in a social setting. These patterns of behavior begin in late childhood and persist through adulthood. The individual is usually only slightly aware of the effect of his behavior on others. Their behavior often conspires to reduce their chances at social success. The personality disorders are quite common. Psychotherapy is of minimal help and medications are of little help. Probably the most famous personality disorder is *antisocial personality disorder* where there is a pervasive lack of empathy and respect for others persons and property starting at a very early age. These individuals rarely perform adequately in school and often drop out early. They get involved in minor criminal offenses and often have long histories of interaction with the legal system. They are usually superficially friendly, manipulative, and unable to learn much from repeated mistakes. They represent a disproportionate percentage of the prison population. Cognition seems mildly impaired with deficits in the ability to understand cause of effect as well as limited ability to emotionally attach to others.

Substance Abuse

This broad behavioral phenotype involves continued and excessive use of addicting substances with resultant problems involving health, relationships, or employment. The most typical example is *alcoholism*. The disorder is quite common with exact prevalence dependent upon the looseness of the diagnostic criteria. It is not uncommon to have affected individuals complain that their substance use is an attempt to "treat" anxiety or moodiness.

IMPLICATIONS OF THE LIMITED NUMBER OF BEHAVIORAL PHENOTYPES

Ninety percent of all psychiatric problems fall into these ten broad phenotypes. There are less than a hundred other specific psychiatric disorders, but together they account for only about ten percent of all psychiatric problems. With the exception of mental retardation, very little progress has been made into subtyping these phenotypes in any etiologically meaningful way. In fact, based on observable behavior alone, it has proven to be virtually impossible to subtype them further. Since behavior cannot be used to subdivide, it is a logical question to ask if there is any other level of organization that can be used to subdivide. To date, no attempt at subdividing using neuropathology, histology, neuroimaging, or laboratory markers has proven to be very helpful except in the case of mental retardation and in a few specific examples of other disorders. The absence of progress suggests that the answer to the question of etiologic subtyping is to look to even deeper levels of organization, namely to the information in the genetic code.

What is the implication of the limited number of phenotypes on the genetics of behavior? The first thing that must be realized is that clinical homogeneity does not reflect etiologic homogeneity. No plausible argument can be made that disorders with prevalences of greater than one percent could be due to a single genetic cause. In mental retardation where the most subtyping has been done, there are literally hundreds of specific genetic and nongenetic causes of mental retardation syndromes, even though most causes of mental retardation remain unknown. Using mental retardation as a model, it appears that when genetic analysis of sufficient clarity is found there will be a very large number of specific genetic causes of all of these broad behavioral phenotypes.

The second implication is that if a variety of problems all lead to the same phenotype, as in mental retardation, then it implies that there are a limited number of ways that the brain can demonstrate abnormality. With all of its complexity, specific brain defects seem to be funneled into relatively few patterns of behavioral abnormality.

Chapter 2

Basic Principles of Genetics and Their Implications for the Genetics of Behavior

BASIC GENETIC CONCEPTS

The observation that "like begets like" has its origins in antiquity. Ancient philosophers puzzled over the phenomenon, while ancient herders and farmers used it to improve the attributes of domesticated animals and plants. Throughout the classical and middle ages man continued to use this empirical knowledge to improve yields and stock with an increasing awareness that certain traits could be "handed down" from one generation to another. Yet, it was not until the mid 17th century that science began to decipher the elaborate mechanism of inheritance. It was during this early phase that Harvey postulated that living organisms arose from eggs(1651), Hooke first described cells(1665), and Leeuwenhoek observed sperm in the semen of man and animals(1677). The natural progression of this work led to the generalization by Spallanzani in 1769 that the exact method of fertilization was the contact of sperm from semen and eggs from the female. This seemingly simple observation helped "demystify" the reproductive process, and accelerated questions regarding the inheritance of characteristics(Jacob, 1973; Stubbe, 1972).

The next century brought intense interest to the question of inheritance of traits. Many believed the process to be a blending, where both traits molded into each other. For instance a tall plant crossed with a small plant would yield a medium size plant. However, careful observation often showed exceptions to this belief. The Augustinian monk Gregor Mendel performed a series of experiments with the garden pea in the 1850's and 1860's and published his classic paper *Experiments in Plant-Hybridization*, which essentially proved that some traits were inherited as discrete entities and not simply molded together. This important discovery was ignored for thirty-five years(Mendel, 1865).

The first decade of the 20th century marked the beginning of modern genetics. Within a few years Mendel's laws of inheritance were rediscovered and W.S. Sutton postulated "the chromosomal theory of inheritance." This theory, simply stated, postulated that sperm and egg were single cells that contained structures named *chromosomes,* and that chromosomes carried the inherited traits (now called *genes*). Furthermore, the observed behavior of the chromosomes was consistent with Mendelian theory. This observation explained two natural processes that had already been observed. The division of cells as they replicated(called mitosis), and the unique division of egg and sperm cells, which halved the number of chromosomes(called meiosis).

The Cell Cycle

In *mitosis* the chromosomes of a cell duplicate before division, thus ensuring that each new cell receives the same number of chromosomes. The behavior of dividing cells has been thoroughly studied. Dividing cells pass through the following steps(called the cell cycle). First, the cell increases in size. Then there is a period of DNA synthesis when the chromosomes are duplicated prior to cell division. The chromosomes then condense and are pulled apart to form separate cells. Many cells function in a suspended state for long periods with no dividing. Mitosis is the mechanism of development. It is how we go from one cell to billions of cells. Errors in mitosis during early development can lead to disease and have implications for behavioral genetic research.

The question of how germ cells replicated without doubling the number of chromosomes was answered by observation of the process of *meiosis*. Meiosis serves as a mechanism to reduce the number of chromosomes in half before they rejoin as egg and sperm to begin the development of a new individual. The 23 pairs of chromosomes(diploid set) are reduced to 23 individual chromosomes(haploid set). Meiosis is a special form of cell division restricted to the germ cells. It has many special properties and consequences. During meiosis there is one round of DNA synthesis for two cell divisions. In the process, there is a period where each chromosome in a pair has been duplicated, but remain physically adjacent to each other (up to 45 years in women, less than 64 days in men). This allows for an interaction between the chromosomes and exchange of material by "crossing over." This important mechanism serves to increase genetic variation. It is estimated that over 50 separate recombinations of genetic material occur this way in the development of each egg and sperm. This genetic variation is no accident. It is an essential part of many genetic processes. Even the distribution of the chromosomes to the gametes(either paternal or maternal) is random and it has been calculated that there are two to the 23rd power or 8.4 million different combinations of parental genes possible for each germ cell.

The Chromosome Number in Man

The chromosome theory was confirmed by the experimental studies of Morgan, Sturtevant, Muller, and Bridges in the period from 1910 to 1916. It was shown that genes were not only present in the substance of the chromosomes, but that they had a predictability that allowed for them to be mapped. These maps, like the chromosomes, were linear. The implication was that traits, all hereditary traits, had to be, in some way, in the chromosomes. It seemed impossible that these tiny amorphous nuclear structures could determine a trait, let alone thousands of traits. Despite this intense interest in chromosomes, such basic questions as the chemical composition and even the number of chromosomes in man would not be answered for a generation. It was until 1956 that Tijo and Levan provided clear evidence that there were 46 chromosomes in human cells(or 23 pairs). These chromosomes are comprised of 22 pairs called "autosomes" and a pair of "sex chromosomes," which are both "X" in females and a "X" and a "Y" in males. Further progress on chromosome morphology came with banding pattern techniques that allowed for the identification of individual chromosomes and classification by length. This advance allowed for the first time for specific behavioral abnormalities to be shown to be causally linked to specific chromosome abnormalities.

DNA: The Molecule of Life

By the 1940's, it was known that chromosomes were supermolecular complexes containing protein and deoxyribonucleic acid or DNA. The debate over which of the two substances carried the genes was answered with the publication of Avery et al.(1944) where the transforming substance was shown to be DNA and not protein. However, this finding initially only served to confuse the issue more. How could a substance with essentially four repeating nucleotides be the gene? The answer came in 1953 with the dramatic presentation of the Watson-Crick model of the DNA molecule. With this model, it became clear that DNA had all the properties necessary to be the "molecule of life." DNA has several attributes which explain its function. First, the hydrogen bonds between the bases can easily come apart. This was immediately seen as the mechanism, somewhat analogous to a zipper, by which chromosomes separated during replication. The copy is easily made because each base has only one analogous base. The entire message is determined by the sequence of DNA "base pairs," labeled A for adenine, G for guanine, C for cytosine, and T for thymine. It is estimated that there are three billion of these base pairs in the total amount of human DNA (genome). By this time, genes were no longer seen as abstract "traits," but instead were recognized as sequences of DNA, which provided the information necessary to construct specific protein molecules. This was expressed as the famous *one gene-one enzyme* hypothesis of Beadle and Tatum. How is the information in DNA translated into the protein molecule? The answer was worked out in the famous *central dogma theory* coined by Francis Crick. The information from DNA is *transcribed* to a chemical known as messenger RNA, which is then *translated* to protein.

The Genetic Code

How does the information in DNA ultimately determine the production of a protein molecule? It is done with the use of a code, a "genetic code." Since DNA has only four base pairs and there are some twenty amino acids(the building blocks of protein), it was speculated that only sets of DNA base pairs could have the specificity to code. For instance, using only one base pair would give four possible codes A,C,G,T. Using two base pairs would give eight AA, AC, AG, AT, CC, CG, CT, CT, GG, GT. Only with three base pair combinations (64 possible) would there be enough specificity. However, since only a couple dozen codes are needed with 64 possibilities, there is a lot of redundancy. The following is a list of the 20 amino acids and the genetic code responsible for each.

alanine	ala	leucine	leu
arginine	arg	lysine	lys
asparagine	asn	methionine	met
aspartic acid	asp	phenylalanine	phe
cysteine	cys	proline	pro
glutamine	gln	serine	ser
glutamic acid	glu	threonine	thr
glycine	gly	tryptophan	trp
histidine	his	tyrosine	tyr
isoleucine	ile	valine	val

First Base	***Second Base***				***Third Base***
	U	**C**	**A**	**G**	
U	UUU phe	UCU ser	UAU tyr	UGU cys	**U**
	UUC phe	UCC ser	UAC tyr	UGC cys	**C**
	UUA leu	UCA ser	UAA stop	UGA stop	**A**
	UUG leu	UCG ser	UAG stop	UGG trp	**G**
C	CUU leu	CCU pro	CAU his	UGU arg	**U**
	CUC leu	CCC pro	CAC his	UGC arg	**C**
	CUA leu	CCA pro	CAA gln	UGA arg	**A**
	CUG leu	CCG pro	CAG gln	UGG arg	**G**

A	AUU ile	ACU thr	AAU asn	CGU ser	**U**
	AUC ile	ACC thr	AAC asn	CGC ser	**C**
	AUA ile	ACA thr	AAA lys	CGA arg	**A**
	AUG met	ACG thr	AAG lys	CGG arg	**G**
G	GUU val	GCU ala	GAU asp	GGU gly	**U**
	GUC val	GCC ala	GAC asp	GGC gly	**C**
	GUA val	GCA ala	GAA glu	GGA gly	**A**
	GUG val	GCG ala	GAG glu	GGG gly	**G**

The implication for behavior from the elucidation of the chromosome theory and the genetic code is the further rejection of vitalism and a reinforcement of the notion that understandable rules underlie all complex situations including the genetics of behavior.

Mutation

It is estimated that the human genome contains 50,000 - 100,000 genes. With this quantity of information, reproduction must be a very precise mechanism. However, errors do occur, and they are known as *mutations* . A mutation is a change in the DNA from the previous generation. Herman Muller received the Nobel Prize for showing that, among other causes, x-ray's induce mutations. The changes can involve single base pairs, or the deletion or rearranged of whole sections of DNA. One of the most illustrative examples of a mutation is sickle cell anemia. In this illness, a single base pair out of hundreds of base pairs in the hemoglobin molecule is altered, but this leads to a different structure of the hemoglobin molecule, which then combines with other hemoglobin molecules to distort the red blood cell into it's characteristic sickle shape. The defective RBC then leads to decreased oxygen carrying capacity and clogging of the small arterioles leading to the clinical symptoms of the disease. The concept of mutation is very important in genetics(or psychiatric genetics) because it is the mechanism by which genetic illness is introduced.

Darwin's theory of natural selection, the initial theory of evolution, predicted that small, incremental differences in organisms are selected for and ultimately become over represented in future generations leading to evolutionary changes including new species. By the turn of the century, however, biologists were increasingly skeptical that selection of continuously variable traits could lead to new species. For instance, when populations were selected for some trait such as height, what was found was that the average height of the subsequent generations did increase, but not more than the tallest organisms within the original variation. Using size calculations on tens of thousands of beans, the Dane Wilhelm Johannsen showed, in 1909, that evolution could not proceed with selection alone. It was Hugo de Vries's mutation theory that solved the problem by allowing for discontinuous variation by the rare presence of "sports" or what are now called mutations. He had noticed that in large groups of organisms, there would occasionally be an odd new type. Occasionally these new types exceeded the preexisting variability of the group. It is this rare mutation that is the actual basis for evolution. This is not to say that mutations always lead to evolutionary changes. The vast majority of mutations are so deleterious to normal functioning that they lead to the death of the individual or to illness. Occasionally a mutation provides some selective advantage and then becomes increasingly selected for leading to evolutionary change.

Chromosomal Organization and Gene Structure

The heredity material is a macromolecule with information stored in the sequence of base pairs. However, evidence is accumulating that only a relatively small portion of the DNA sequence is for so called structural genes. Structural genes lead to the production of protein. There are an estimated 50,000 structural genes with an average size of approximately 5,000bp which then accounts for only 250,000,000 of the estimated three billion base pairs. What is the rest of the DNA for? Some of the DNA is so called repetitive sequences that are repeated thousands of times. The function is unknown. The Alu repeat, for instance, contains over 300,000 copies of the same 300bp sequence. Certainly, this DNA is not "junk" and plays some important role in gene regulation, chromosomal architecture or chromosomal replication. Until 1977, it was thought that genes are single sequences of DNA that are coded into RNA and then to protein. However, further study has shown greater complexity. It is now known that there are pieces of DNA within a gene that are not translated into protein. These intervening sequences or "introns" are somewhat of a mystery, but appear to be a very common phenomenon. The DNA segments that are translated into protein are called "exons."

GENETIC FACTORS IN HUMAN ILLNESS: MENDELIAN PATTERNS OF INHERITANCE

Each chromosome in a pair is referred to as a *homologous chromosome*. Each homologous chromosome has a separate copy of each gene, so there are two copies of each gene. These two copies are often slightly different in their base pair composition, even though they code for the same genes. These two genes (or *loci*) are called *alleles*. This is a central concept in the inheritance of abnormal traits. If a disease is due to a genetic cause, it is because at least one allele is defective. Since there are two copies of each gene (one from the sperm and one from the egg) often there is one abnormal and one normal gene. Many times the body can tolerate this. The individual is referred to as *heterozygous* for the disease trait. If two abnormal genes are present then the individual is homozygous for the abnormal gene and disease usually is the result. Because of the nature of genetic transmission, certain patterns emerge through generations that can indicate the exact type of inheritance that is involved.

Dominant Inheritance

Dominant inheritance was first reported by Mendel in garden peas. In dominant inheritance, one of the affected person's two alleles is defective. The parent then has a 50 percent chance of passing that disease gene to an individual child. However, as a consequence, any child that does not receive the gene is free from the illness except for the extremely rare chance of a new mutation. In pedigrees with fully penetrant dominant transmission of a trait one finds an affected member in each generation without skipping generations. On average about 50 percent of at risk offspring will have the condition. One of the complicating factors in humans for dominant transmission is that some disorders, especially behavioral disorders, do not fully develop until later in life. If a person is assessed before the onset of the disorder, it will have appeared to have skipped a generation. This also has implications for research, which will be discussed later. Another complicating factor is the concept of *penetrance*. All individuals with dominant traits do not necessarily express the trait phenotypically, when this happens it is called reduced penetrance. In general, dominantly transmitted diseases tend to involve structural proteins. An example of a psychiatric genetic disease transmitted in this fashion is Huntington chorea.

Recessive Inheritance

Recessive inheritance requires that both alleles be dysfunctional in some way. They do not have to be mutated in the same manner, but both copies must be abnormal in some way. For an individual to get two copies of an abnormal gene, it is required that one gene come from each parent(with the exception of the extremely rare uniparental disomy phenomenon). Each of these parents are considered heterozygous, in that they each have one normal and one abnormal gene. It is of course possible to inherit an abnormal gene from a parent who is homozygous for the abnormal gene and is affected, but it is rare. On the average, 25 percent of offspring will be affected with the recessive disorder. Most recessive genes are considered to be enzymes and cell membrane proteins. An example of a psychiatric genetic disorder transmitted by recessive transmission is phenylketonuria.

Pedigrees segregating a recessive trait will reveal skipped generations, a sporadic appearance to the affected individuals, and most importantly consanguinity (intermarriage). The odds of rare recessive traits combining are enhanced tremendously with the presence of a common ancestor. The consensus is that the rarer the condition, the higher the likelihood of consanguinity. Since the odds of an individual being affected when both parents are heterozygotes is 1/4. Many families that might have had affected children will not show the disorder. This can be expressed as:

$$(3/4)^n$$

Family size is "n." Small family sizes will limit the ability to see the disorder, perhaps for generations. In addition, unaffected siblings of such families have a high likelihood of being a carrier (a 2/3 chance). There are consequences for behavioral disorders due to recessive traits. Most behavioral disorders, for instance, are associated with decreased reproductive fitness, which further frustrates the recognition of a recessive inheritance pattern.

X-Linked Inheritance

Disorders due to genes on the x-chromosome reveal a fairly distinct pattern of transmission. Since males have only one X chromosome they have what is called a *hemizygous* state in that they have only one copy of the gene. When the gene is abnormal then they will be affected. Females have two copies of the gene and unless it is a rare x-linked dominant trait they will be carriers only. Thus, pedigrees will show a high excess of affected males with no father to son transmission. This phenomenon of hemizygosity in males has been used as an explanation for the greater number of mentally retarded males than females.

GENETIC FACTORS IN HUMAN ILLNESS: NON - MENDELIAN PATTERNS OF INHERITANCE

The Concept of Predisposition to Disease

The idea that individuals have different inherent risks for incurring certain diseases is an ancient concept(Garrod, 1931). A number of different terms have been used and the precision of those terms have varied. However, the term diathesis, which originated in the French school of the 18th century, became the most widely accepted description of the concept until replaced by "predisposition" and "constitutional factors." However, the clinical observation that individuals are born in families with predisposition to an illness, even if seen for several generations did not provide the predictive ability and logical basis that Mendelism provided.

Simple versus Complex Diseases

Since shortly after it was first recognized that some apparent genetic disorders had patterns of transmission that did not fit easy Mendelian patterns, there was a need for a category of "complex" genetic disease. Factors such as reduced penetrance, heterogeneity, and pleiotropy have long been suspected to play a role in the transmission of complex genetic traits. Diseases are considered "multifactorial" when genetic and non-genetic causes are involved in their transmission. For instance, alcoholism has been shown to have some genetic basis. However, it also requires the presence and social acceptance of the use of alcohol. Alcoholism then is, by definition, a multifactorial disorder. In contrast, a *polygenic disorder* is a genetic condition that is due to the interaction of several genes, each of which alone is insufficient to cause the disease phenotype, while the small effects of each gene can in total lead to a disease phenotype. In addition to these factors, there are a number of other factors that serve to increase phenotypic variation and obscure single gene influences.

Allelic heterogeneity occurs when alleles of one gene give rise to different phenotypes depending upon the nature of the gene abnormality. Duchenne muscular dystrophy provides an illustrative example. For many years clinicians suspected a link between the more severe Duchenne muscular dystrophy and a milder form of muscular dystrophy known as Becker muscular dystrophy. The molecular basis of this link was clarified when both conditions were due to different mutations in the same dystrophin gene. Duchenne was caused by the loss of one portion of the gene while Becker was caused by the loss of another portion of the gene. Allelic heterogeneity has proven to be of some importance in virtually every gene that has been sequenced and studied.

Heterogeneity is the concept that there are several genes, each of which when abnormal can all lead to a similar phenotype. This phenomenon is well documented in human genetics and serves to increase variation. For example, two individuals with disease X may marry and it might be predicted that 100 percent of offspring would be ill with a recessive trait. The results, however, show that no children are ill. The reason is that each parent had the disease due to different genes. Each child will be a carrier for both genes, but none will be ill. It is quite likely that this type of heterogeneity will be highly important in the genetics of the broad behavioral phenotypes.

Pleiotropy is the phenomenon where a single gene gives rise to a number of effects, some of them seemingly unrelated to each other. This source of variation is of growing importance in understanding the genetics of behavior. For instance, in schizophrenia there is some evidence that eye tracking disorders and psychosis may be pleiotropic manifestations of the same genetic illness.

Dominance is when two allelic genes interact in a way that is not predicted by either gene alone. This may contribute to severe phenotypes or may even mask illness in some cases.

Epistasis is the interaction between different loci. This is a common source of variation, but is difficult to clearly demonstrate. The assumption is that complex subtle effects of other loci than the one in question leads to significant differences in phenotypic expression.

Mitochondrial DNA. Since the transmission of the mtDNA molecule (16,569 nucleotide pair) is strictly by maternal inheritance and involves a variable number of mtDNA molecules at a time, this is a potential source of variability. Specific diseases are known to be caused by mtDNA mutations and they include effects on the CNS (Wallace, 1995).

Developmental plasticity is another source of phenotypic variation, and may prove to be one of the most important. Variation occurs even in totally inbreed lines of research animals. These animals are all genetically the same. The variation occurs even when the environment is fixed. It appears that this variation is due to random events in development that interact in unpredictable ways. Schizophrenia spectrum disorders, for instance, may be due to developmental plasticity in the manifestation of the same gene in different individuals.

How to Explain the Genetic Basis of Low Recurrence Risk Diseases

One of the most debated problems of psychiatric genetics is the presence of disorders that appear familial and may even have high heritabilities in twin studies, but have recurrence risks in the low range of 1-15 percent.

Multifactorial Threshold Model - This model was proposed by Penrose(1953) and others (Edwards, 1960; Carter, 1965; Falconer, 1965) to explain the genetics of disorders with very low recurrence risks (much less than 25%). These are the disorders that cannot be explained by Mendelian transmission without invoking "reduced penetrance" or another modifying factor. The model assumes an unspecified admixture of genetic and environmental risks that create a smooth curve of liability that only causes the disease in question when the liability exceeds a theoretical threshold. The genetic component is postulated to be polygenic in nature because the risk is seen as continuous not discontinuous as would be seen in a disorder with a strong single gene effect. This model has had wide acceptance, although rigorous experimental evidence for its existence is not strong.

Developmental Plasticity, Stochastic, or Randomness Models- Ventricular septal heart defects have a low familial recurrence rate, but familiality is one of the strongest and most consistent etiologic factors. Newman (1985) studied the factors involved in development of VSD's and speculated that the epidemiologic findings of VSD could be consistent with a randomness factor that modified a genetic factor. Kurnit, et al. (1987) expanded on this idea by using computer simulation to suggest that a single gene whose phenotypic effects were modified by developmental randomness could produce the same sort of continuous liability risk as seen in multifactorial threshold models. By simulating endocardial cushion outgrowth, they were able to demonstrate different phenotypic outcomes using identical initial parameters. In addition, by varying the value of the "adhesiveness" parameter in their model they could mimic a shifting liability curve with more individuals theoretically effected as the parameter decreased.

What smooths out the curve from the discontinuous curve normally seen in Mendelian traits is the large role of chance. The authors state that initial random cell movements occur "thereby amplifying the effects of initial stochastic variations and introducing a significant role for chance." They term this variation "stochastic variation," and point out that it can smooth the liability curve without other genetic or environmental influences. Recurrence risks from this model are similar to those predicted by James (1971).

Experimental evidence of randomness factors influencing final phenotypic outcome are growing. It was noted half a century ago (Kartagener, 1933; Feldman, 1935) that simple chance determined whether individuals homozygous for the situs inversus gene would have situs or not (50% by chance stayed normal while 50% had situs). Therefore, only 12.5% of "at risk" individuals had the final phenotype of situs. Retinoblastoma is a more recent example of chance events (i.e. the second mutation) having phenotypic effects (Cavenee et al., 1983).

How to Explain Familiality in Very Rare Disorders

Some clinical disorders appear to occur with great rarity. These disorders usually happen only once in a family, but occasionally they may occur twice or more in a given family. This has often led previous investigators to speculate the disorders are recessive. However, lack of consanguinity and lack of multiple affected family members speaks against recessive inheritance in most of these disorders. A more likely scenario is that the disorders are due to germ line mutations. In this scenario, the parents are unaffected, but there exists a new mutation in the germ line that arose since conception of the parent. The percentage of germ cells affected would depend on when the mutation occurred. In multiple affected families it could be argued that it was an early mutation affected large numbers of germ cells, whereas in sporadic families in may have only affected one individual. The disorder itself is usually of a severity that no further reproduction occurs with the affected individual.

SPECIAL TOPICS IN GENETICS WITH RELEVANCE FOR BEHAVIOR

Why is There Sex?

One of the simplest, yet most vexing questions of the geneticist, is why is there sex? Many organisms are not sexual, they simple duplicate like bacteria, or develop by parthenogenesis. Why all the complexity of sex? The answer lies in the need for diversity. Animals that reproduce without sex, have the advantage of simplicity, but pay the price in uniformity. Each offspring is a clone of the parent (except the rare case of mutation). This limits the organism's response to the environment. In times of great environmental change, there may not be enough diversity in the population to survive. Sexual reproducing animals have the advantage of diversity. Each time a new individual is created through sexual combination of two genomes, including the admixture and recombination of genes, an infinite diversity of genotypes is possible. This creates a population of individuals with tremendous diversity, which is a payoff in times of environmental change. Extinction is less likely, mutations are recombined in ways that simply would not happen in nonsexual beings.

In terms of behavior, this diversity also leads to greater behavioral ranges. The message is simple. The greater the behavioral diversity in a set of organisms the greater the survivability. However, there is a cost of the greater diversity of behavior. Many more combinations of behavior are going to be formed than fit the existing environment. Many of these diverse forms of behavior are, by chance, going to be poorly suited to their environment. This concept is fundamental to understanding pathology. When the range of behavior is increased, the likelihood is that some extremes of that range will be ill suited for adaptation.

Polymorphism and Biological Individuality

Multiple genes at the same allele are termed *polymorphism.* In the first decades of genetics the idea that there could be more than two genes for a given allele was only theoretical. It is now known that polymorphism is almost the rule. The ABO blood system is an example of a polymorphic system with three genes A, B, and O. Individual genotypes are either AA, AB, AO, BB, BO, or OO. The whole population can be divided into these six classes. It is easy to see how this type of polymorphism combined with other polymorphic systems generate infinite possibilities of combinations to the point where no two individuals could have the same alleles for every gene. Even before this was understood genetically, Robert Williams studied individual uniqueness in terms of biochemistry and came to the conclusion that subtle variation at biochemical and morphological levels reflected a broad underlying genetic diversity.

Peter Medawar used the study of immunocompatibility to reach the same conclusion, namely that there is so much genetic polymorphism that every individual is a unique combination of genes that will never be duplicated. Simply stated all individuals are unique. It is usually also stated as a caveat that identical births are an exception to this rule since they are derived from the same fertilized egg. However, somatic genetic changes, even if minor in comparison to the genetic differences between nonidentical births still ensure that even twins are genetically nonidentical. A point of great significance as will be discussed later.

The implications of genetic individuality for the understanding of behavior is immense. Just as we have different eye colors, different enzyme levels, or morphological variations, we must have different brains. If the genetic code itself is unique, then that uniqueness is amplified throughout development. At every step of development there must be ramifications of the difference in the genetic code including subtle effects on embryonic cell migration and cell death patterns. The result is that the end product, the functioning human brain, is entirely unique. If no two brains are alike, then no two sets of behavior can be alike even before any learning or other environmental effect takes place.

Anticipation

Of special interest to those interested in the genetics of behavior is the concept of anticipation. Historically this concept goes back to the idea of "degeneration" first espoused by Morel in 1859. This concept, which was popular for the next fifty years, stated that there was a general worsening of the behavioral problems each generation leading to extinction of the problem. In its classic form it was believed that the first generation might exhibit mild neurosis, the second psychosis, and the third imbecility and the fourth early death. This concept fell into great disfavor early in the twentieth century and was seen as fatalistic, incorrect, or at best an ascertainment bias. Geneticists renewed the concept in the form of "anticipation." Anticipation describes a clinical phenomenon where there is increasing severity of disease symptoms or earlier age of onset in members of succeeding generations. Again this concept was virtually ignored as a form of ascertainment bias. However, recent findings of unstable triplet repeats in fragile X syndrome and myotonic dystrophy have brilliantly legitimized the concept. In these conditions, the number of triplet repeats are expanded(or rarely contracted) in successive generations, with concomitant increase in severity and earlier onset, confirming a molecular mechanism of anticipation. Perhaps the most important aspect of this finding is the legitimization of clinical observation as an important tool in understanding behavior. There is a tendency to consider behavior unsuited for clinical observation, and that only biologic markers or neuroimaging have the precision necessary to determine anything of value. The clinical observation of anticipation speaks against this trend in thought.

Hardy Weinberg Equilibrium

The Hardy-Weinberg equilibrium refers to an algebraic formulation of gene frequencies in a static population. In a two allele system, if a given allele has a frequency of p then the other allele must have a frequency of 1-p or q. Or in other words p + q = 1. This relationship can then be made into a binomial expansion giving the following:

$$p^2 + 2pq + q^2 = 1$$

Given a two allele system A and a. There are three possible genotypes AA, Aa, and aa. These genotypes correspond to the formula above. The frequency of the AA genotype is p^2, the Aa genotype is 2pq, and the aa genotype is q^2. This formula can be very helpful in predicting gene frequencies for recessive traits. For instance if a recessive disorder (aa) has an incidence of 1 in 360,000 births, then q^2 equals 360,000 and q is the square root or 1/600. Therefore, we know that the frequency of the disease gene in the general population is 0.002 and the normal gene frequency is p = 0.998. The importance of the Hardy-Weinberg equilibrium for behavior is immense. In stable populations, it predicts that little changes can be expected in disease gene frequencies unless large selective pressures are exerted. Attempts at eliminating recessive traits by asking the affected persons not to have children would be of little use, since the carrier frequency is 1 in 250 individuals versus the affected frequency of 1 in 360,000 individuals. Eliminating reproduction in the affected would not even dent the huge reserve of carriers. The implication is that certain levels of genetic abnormality are unavoidable in stable populations and are the price to be paid for variation and evolution.

The Question of Inheritance of Acquired Traits

The belief that traits acquired by an individual during his or her lifetime could be passed on to the next generation through heredity has its origins in antiquity. Physical as well as mental traits were all thought to be transmitted. It was not until the end of the 19th century that the concept began to be dispelled. August Weissmann argued strongly against acquired characteristics in his work *The Germ Plasm: A Theory of Heredity* (1893). Based on the weight of his arguments and others that followed, the concept of the inheritance of any trait physical or mental has been virtually totally abandoned. If there is any place where such concepts still linger, it is in the field of behavior and mental illness. There are still vestiges of Lamarckianism in behavior theory. A classic example is the Jungian concept of collective unconscious. According to Jung, there are symbols that have become ingrained in man's unconscious, that are not learned culturally, that have become that way due to repeated exposure to them by our ancestors. This is a clearly Lamarckian idea that has no credible scientific basis.

Assortative Mating

Assortative mating is non-random mate selection and is a well-documented event in human populations. Any model that assumes random mate selection will be wrong. As long ago as 1903 investigators found evidence of positive correlations between mates for traits such as age, stature, and intelligence. Physical correlations in human populations include .75 for age and .20 for IQ correlation. The implication for behavioral genetics is that some forms of psychopathology seem to exhibit assortative mating. Individuals with mood disorders seem to have a much higher incidence of marrying someone with a mood disorder than is predicted by chance. This is particularly problematic in pedigree and linkage studies where the introduction of new disease genes outside of the main pedigree line can cause great problems in analysis.

Parent of Origin Effects

Differences in affection status due to which parent the individual inherits a disease gene or chromosomal effect from is an idea that causes consternation in classical genetics. This clearly non-Mendelian form of inheritance has been difficult to explain, but appears to occur in a number of disorders. Two hypotheses have been offered to explain this finding.

Differential Germ Cell Line Formation - One explanation for differences in maternal and paternal derived gametes is the number of cell divisions required to form the respective gametes. Male spermatogenesis is an ongoing event where hundreds of cell divisions may have occurred by the time the sperm used at conception is formed. Female oogenesis is delayed and averages about 24 cell divisions. Mutation rate is correlated with cell division and cell replication, which may explain some cases of paternal parent of origin predominates. In other words, mutation may be more likely in paternal germ cells than maternal germ cells.

Imprinting - Involves the inactivation of an allele differentially due to passage through either spermatogenesis or oogenesis. This finding is poorly understood and may involve methylation patterns of DNA.

Chapter 3

The Behavioral Consequences of Chromosome Abnormalities

NOMENCLATURE

Appreciation of the role of chromosome abnormalities in the production of physical and behavioral problems depended first on the general acceptance of the role of chromosomes in heredity. The role of chromosomes as the bearer of inherited traits was generally defined in the period 1903-1916 by Sutton, Morgan and others. Another problem was to accurately count and classify the number of human chromosomes. While it was known that *Drosophila* had eight chromosomes from early in the twentieth century, the accurate number of human chromosomes (46) was not established until 1956 by Tijo and Levan. With the accurate number of human chromosomes determined and with improved techniques for classifying chromosomes, the field of human cytogenetics began. The technique for visualizing human chromosomes involves isolating the white blood cells of an individual, making a cell culture, and chemical treatment of the cells. A smear of metaphase chromosomes from a single cell can be obtained and this smear is then photographed and the chromosomes cut out and arranged by size and the location of their centromere (connection point) into 23 haploid pairs (including the paired X and Y chromosomes). This test is commonly referred to as a "karyotype."

Very early in the study of human chromosomes it was clear that a standard nomenclature was needed to describe the various deviations from normal that were being seen. In 1960, a meeting was held in Denver, Colorado that produced the so-called Denver

Classification. This classification has since evolved and can be quite complicated. However, the basic principles are simple. The 23 chromosome pairs are numbered from 1 to 22 with the largest being number one and the smallest number twenty-two (the X and Y chromosome making the twenty-third pair). A nomenclature for describing an individuals chromosome complement was also described. The first number refers to the total number of chromosomes (for a normal male or female - 46). The next reference is the status of the sex chromosomes (male XY, female XX). So a normal male would be referenced as 46,XY, and a normal female as 46,XX. The vast majority of all individuals ever tested have been one of these two karyotypes, however, almost as soon as the testing began there were abnormalities detected. The following is a review of the behavioral consequences of specific types of chromosomal abnormalities.

BEHAVIORAL CONSEQUENCES OF THE LOSS OF A CHROMOSOME

The consequence of the loss of a chromosome in the developing organism is evidently catastrophic to the fetus. Very few examples of live births with an absence of one of the 22 autosomes exist. Apparently, spontaneous abortion is the rule for such fetuses. An argument can be made that the large number of genes on even the smallest chromosomes are essential to viability and the fetus cannot develop without them. There is one notable exception to the idea that the absence of a chromosome is inconsistent with viability, that is the condition known as 45,XO or Turner Syndrome. Individuals with this disorder have one X chromosome, but lack a second sex chromosome.

45,XO (Turner Syndrome)

Individuals with Turner syndrome are phenotypically female. Physical findings include short stature, low posterior hairline, streak gonads, webbed neck, and cubitus valgus. Turner syndrome occurs in one in 3,000 female births. Behavioral aspects of this disorder have only recently been studied and preliminary results suggest that cognitive effects are mild with mental retardation only very slightly over represented, if at all. Other behavioral consequences appear to be more subtle, with studies suggesting problems with visual-spatial ability resulting in learning disorders. This "spatial agnosia," as it has been called, seems to impair right-left orientation and arithmetic skills (Nielsen, et al., 1981). Christensen et al. (1981) reported evidence based on neuropsychologic testing that Turner women have abnormalities in the right hemisphere and frontal brain area. They found little evidence for a diffuse cerebral abnormality. A few studies have reported some improvement in spatial ability with the administration of estrogen. (Nyborg et al., 1981) The syndrome seems to have little impact on sexual orientation since homosexuality is not over represented in this group.

All things considered, for missing a whole chromosome, there is not much abnormality in people with Turner syndrome, especially considering the consequences of the loss of other chromosomes. There are several reasons why this is the case. Only about one in 40 conceptions with Turner syndrome are not spontaneously aborted. This process selects for only the most viable and, therefore, the least affected fetuses. Also, the Lyon hypothesis of X inactivation predicts that even in normal XX karyotypes one of the X chromosomes is inactivated early in development. This leaves every female with something close to an XO genotype. There is no viable YO syndrome, and a minimum one X chromosome must be present for viability.

BEHAVIORAL CONSEQUENCES OF THE ADDITION OF A CHROMOSOME

The addition of a chromosome generally leads to non-viability. It is not immediately obvious why having three copies of a chromosome would cause any damage since the only difference would be the presence of three copies of some genes. The fact that extra chromosomes do lead to such problems suggests that there is an intricate balancing act in development that can be offset by the slightest deviation. There are a number of well recognized trisomy syndromes, but far fewer than the number of autosomes, again suggesting that certain trisomy syndromes are not consistent with life. The following are descriptions of the behavioral consequences of the various trisomy syndromes.

47,XY,8+/46,XY (Trisomy 8 Mosaicism)

This trisomy syndrome was first reported by Caspersson et al. and is extremely rare. It appears to only occur as a mosaic. Physical findings include distinct facies with bulbous nose, abnormal ears, and deep plantar creases. Some brain abnormalities such as agenesis of the corpus callosum have also been noted. Behavioral consequences include near universal mild to moderate mental retardation, although some cases have near normal intelligence.

47,XX,13+ (Trisomy 13, Patau Syndrome)

This trisomy was first described by Patau in 1960. It occurs in one in 5,000 births, and results in cleft lip and palate, hand and finger abnormalities, microphthalmia, plus multiple other congenital abnormalities. Life span in measured in months and severe mental retardation is always present.

47,XY,18+ (Edwards Syndrome)

First described in 1960 by Edwards et al. This disorder occurs in about one in every 3,000 births. Affected individuals usually have growth retardation, hypertonicity, malformed ears, micrognathism, clenched fists with characteristic second finger over third, ventricular septal defect, and small palpebral fissures. Behaviorally, all are severely retarded. Over seventy five percent die within the first six months of life.

47,XX,21+ (Trisomy 21, Down Syndrome)

This is the best studied of the chromosome abnormalities mostly due to its very high frequency. Down syndrome is present in 1 in 600 births. Risk is strikingly correlated with increasing maternal age, however most cases are born to younger women since most babies are born to younger women. Nondisjunction is felt to be the cause. Ninety five percent of affected individuals have trisomy 21 the other five percent have mosaicism or translocation.

Mortality is greatly increased, although this is improving with more aggressive treatment. Physical characteristics include hypotonia, brachycephaly, epicanthal folds, low set and abnormal ears, short neck, short stature, large tongue, simian lines, and fifth finger clinodactyly. Complications include ventricular and septal defects, duodenal atresia, tracheoesophageal fistula, leukemia, and Hirschsprung disease.

Behaviorally, there is almost always mental retardation (mean IQ 30-50). Trisomy 21 accounts for a significant portion of all mental retardation. Down syndrome accounts for 15-20 percent of all institutionalized patients. An interesting finding is that almost all individuals with Down syndrome who live to age 35 show neuropathologic changes consistent with Alzheimer disease.

47, XXY (Klinefelter Syndrome)

First described in 1942 by Klinefelter et al., this disorder is characterized by a number of discernible physical attributes, which makes the diagnosis somewhat probable in the 1 in 500 males affected by the disorder. Principal physical findings include male phenotype, tall stature, gynecomastia, hypogonadism, obesity as adults, and infertility. Behavioral problems have been noted in Klinefelter syndrome patients from very early in the clinical delineation. They have about a one standard deviation difference in IQ compared to siblings (Ratcliffe et al., 1981, Sorensen et al., 1981). However, this appears to be a shift in the normal intelligence curve so that there are still XXY men represented at every level of intelligence. XXY men seem to be prone to dysnomia and other speech and articulation disorders. They have been noted to have persistent immaturity often similar to normal adolescent behavior, but usually more intense. Also they tend to be boastful, but shy. Problems with gender identification have also been reported (Theilgaard, 1981).

Some have questioned whether many of the personality traits detected in XXY males are simply reactions to their body habitus and other physical problems. This question was partially answered by Nielsen et al. in 1969 when they compared hypogonadal males from other causes to Klinefelter males and found that the Klinefelter males had lower IQ' s and the same previously described personality traits. Testosterone replacement has shown some promise in increasing self esteem, although double blind studies are yet to be done. Depressive episodes respond to antidepressants. Although rarer, a number of individual have been found with more than one extra X chromosome. Studies have shown that IQ drops with the addition of each extra X chromosome.

47,XYY ("Supermale Phenotype")

First described by Sandberg et al. in 1961, this disorder has an incidence of approximately 1 in 840 newborn males, making 47, XYY one of the most common genetic disorders in males. It is somewhat ironic, therefore, that the diagnosis is seldom made. This is due in part to a lack of specificity regarding dysmorphology. The physical traits associated with the disorder are mild and highly variable. People with 47, XYY are phenotypically male and have a tendency to tall stature (as much as 5 inches taller than 46,XY on average), accelerated growth in mid-childhood resulting in increased length versus breadth in the cranial vault, feet and hands. The increased height is evident by childhood. The large size is not reflected by particularly good athleticism or strength with many individuals having relatively poor pectoral and shoulder girdle strength. Skin findings include higher rate of nodulocystic acne during adolescence. Occasional abnormalities include increased tendency towards radioulnar synostosis. Additionally they have a slightly increased testicular volume, although fertility is not markedly decreased.

Behaviorally, IQ seems to be affected with a tendency toward borderline intellectual function, although XYY men are distributed throughout the normal intelligence range. The average IQ is about one half to one standard deviation below their unaffected siblings (Ratcliffe, et al., 1981). XXY men seem to have problems with spatial ability, although not as severely as Turner women (Nyborg et al. 1981). Conduct problems do not seem to be elevated as represented by school performance in some studies, but are higher than controls in other studies (Thielgaard, 1981). They do appear to have statistically documented abnormalities in the development of speech leading to an increased rate in the diagnosis of specific learning disorders related to receptive and expressive language.

Some older studies based on skewed samples taken from prisons and juvenile correction centers reported an increased frequency of XYY individuals among populations over representing antisocial personality disorder, but more recent longitudinal studies by Leonard and Sparrow (1986) found no statistical correlation with conduct disorder or antisocial personality. Some anecdotal studies suggest that XYY individuals have quick tempers and other problems with impulsiveness and may be over represented in individuals given the diagnosis of intermittent explosive disorder and other impulse control disorders. Some anecdotal information suggests that these individuals have greater problems with coordination and may be over represented in individuals having developmental coordination disorders. One puzzling aspect of the XYY syndrome is that the Y chromosome has just a few genes. It remains a mystery how an extra copy of these few genes ultimately leads to physical and behavioral manifestations.

47, XXX ("Triple X females," "Triplo - X")

In this syndrome, there is an extra X chromosome. It occurs in about one in a thousand females. Affected females appear to be free of any major congenital malformations and are fertile. However, they do appear to be at slightly higher risk for mental retardation and are over represented in institutions for the mentally retarded. They also seem to be over represented in mental hospitals. Longitudinal, unselected studies like the Edinburgh Newborn Study found 16 of 12,479 (0.13 percent) XXX girls in their sample. These girls had a significant degree of language delay, IQ impairment (mean IQ of 88 versus control group of 118), and social behavior delay (Ratcliffe, et al., 1981).

BEHAVIORAL CONSEQUENCES OF DELETION SYNDROMES

The complete loss of an autosome is apparently incompatible with life. Occasional individuals with mosaic loss of a chromosome have been reported, but this is highly unusual. The only form of chromosomal loss that remains is the so-called deletion syndrome. Early in the study of karyotypes, it was noted that there are individuals with deletions of parts of specific chromosomes. Usually the missing chromosomal material was restricted to one arm or to a small portion of an arm of a chromosome. With improved chromosome banding and visualization, these deletions were charted and syndromes began to be established. This continues to be an area of active research. The following are a selected group of deletion syndromes with known behavioral implications. They are presented in the order of the chromosomes involved.

46,XX, 2q37-

Wilson (1995) has reported four cases of mental retardation, dysmorphic hands and face, and short stature associated with deletions in the long arm of chromosome 2.

46,XY,4p- (Wolf-Hirschhorn Syndrome)

Wolf-Hirschhorn Syndrome (WHS) is associated with a deletion of chromosome 4p and is typified by growth retardation, facial dysmorphology, and mental retardation. The size of the 4p deletion is variable from as much as half of the short arm to cytogenetically undetectable deletions (as small as 2.5Mb). Phenotypically, the size of the deletion does not appear to be critical. Some evidence suggests that the specific region involved is the distal 4p16 region. It is not clear if WHS is due to a single gene deficit, or more likely, a contiguous gene syndrome (a number of distinct genes are lost by one deletion).

46,XX,5p- (Cri -du-chat syndrome)

This abnormality first reported by Lejeune in 1963 has a number of physical and behavioral consequences. Individuals have an unusual vocalization that has been compared to the sound of a cat. Other physical features include microcephaly, down turned palpebral fissures, hyperteleorism, and round face. Behaviorally, all patients with this syndrome have severe mental retardation indicating a severe disruption in normal brain development and function. This syndrome is very likely a contiguous gene syndrome and recently families have been reported with the distinctive cat cry, but with a milder phenotype with only mild developmental delay and smaller 5p deletions (Gersh, et al., 1995).

46,XY,7q11.23- (Williams Syndrome)

Williams syndrome is a disorder characterized by mild mental retardation, gregarious personality, distinctive facies, supravalvular aortic stenosis and idiopathic infantile hypercalcemia. Evidence suggests that it is a contiguous gene syndrome involving deletions on chromosome 7 in the long arm area q11.23. The disorder is associated with loss of an allele for the elastin gene (ELN) located at the deletion site in 96 percent of WS patients (Lowery, et al., 1995).

46, XY, 11q- (Jacobsen Syndrome)

This syndrome is due to loss of the long arm of chromosome 11. Features include mild to moderate mental retardation, dysmorphic head and face, and cardiac defects. Variability is moderate. This condition is rare with an estimated incidence of 1 in 100,000. Life expectancy is limited.

46, XX, 13q- Syndrome

This syndrome involves anomalies of major organ development including abnormal brain development. Severe mental retardation is the resulting behavior.

46,XY,15q11-13 -
(Prader Willi Syndrome and Angelman Syndrome)

Prader Willi Syndrome (PWS) is characterized by mental retardation, hypotonia, hyperphagia (with obesity), and small hands and feet. A prevalence study based on a survey technique from North Dakota revealed a prevalence of about 1 in 16,000 individuals with a 1:1 male to female ratio. Global developmental delay is very common with fifty nine percent mild to moderately retarded (Cassidy and Ledbetter, 1989). Behaviorally, they are often noted to be oppositional, with inability to follow social norms, and with occasional psychotic episodes. Behavioral problems are quite prevalent, often necessitating group home placement. Molecular studies have revealed that approximately 70% of PWS patients have deletions in the paternally derived 15q11-13 region. The other 30% are usually due to maternal uniparental disomy for chromosome 15.

Angelman syndrome is characterized by severe mental retardation, seizures, ataxia, minimal capacity for speech, and a peculiar form of uncontrollable laughter. Molecular studies of Angelman syndrome reveal that maternal 15q11-13 deletions are associated with the phenotype, while paternal uniparental disomy is the norm.

Evidence has been accumulating that imprinting is essential for determining whether the phenotype will be PWS or Angelman Syndrome. Imprinting is a poorly understood process that appears to occur during gametogenesis. The disease phenotype depends upon whether the allele was formed during paternal gametogenesis or maternal gametogenesis. Recent evidence suggests that small nuclear ribonucleoprotein-associated polypeptide (SNRPN) maps to the region involved in PWS and AS. This gene and others close to it have been speculated to be expressed exclusively in the paternal allele in man and mouse. Under this scenario, AS would be due to an excess of paternally derived SNRPN and related products, while PWS would be due to a deficiency in paternally derived SNRPN and related products.

46,XX,16q- Syndrome

This syndrome is associated with mental retardation, microcephaly, minor facial anomalies, and broad halluces.

46,XX, 17p Miller-Dieker Syndrome (MDS)

This syndrome is associated with 17p hemizygosity and results in severe mental retardation. Pathologically, the brain is characterized by lissencephaly and agenesis of corpus callosum (Dobyns).

46,XY, 17p11.2- Smith-Magenis Syndrome (SMS)

First described in 1986, this syndrome is due to a deletion of the 11p region of chromosome 17 (Elsea, et al., 1995). The resulting phenotype consists of multiple congenital malformations and mental retardation. Other features include developmental and growth delay, facial abnormalities, and unusual behavior (Smith, et al., 1986; Greenberg, et al., 1991). Some evidence suggests that the highly conserved enzyme serine hydroxymethyltransferase (SHMT) is hemizygous in SMS.

46,XY, 18q- Syndrome

This syndrome is associated with moderate mental retardation (>95%). It occurs in about 1 in every 40,000 births. Psychotic episodes may also be over represented. (Krag-Olsen, 1981) Preliminary evidence suggests that it is due to a contiguous gene syndrome (Silverman, et al. , 1995). Some preliminary trials of treatment with growth hormone are currently underway, which is hoped will improve cognitive abilities by increasing CNS myelination.

46,XY, 22q11- (Velo-Cardio-Facial Syndrome)

This relatively common deletion syndrome has an incidence of 1 in 5,000 births. The major manifestations are cleft palate and congenital heart defects. However, learning disabilities, psychiatric syndromes, and mental retardation are also associated with the disorder (Pulver, et al., 1994). In addition, a specific late onset psychotic syndrome has bee reported (Shprintzen, et al., 1992). Some evidence from linkage studies have pointed to the area 22q11 as a locus possibly involved in some forms of schizophrenia. Lindsay, et al. (1995) reported on screening 100 patients with schizophrenia for 22q11 probes. Two patients were found, which seems much higher than expected, furthering the hypothesis that this region may account for a small percentage (1-2 percent) of schizophrenia.

46,XY, Xq28- Syndrome

This syndrome consists of developmental delay and speech abnormalities, but has been reported in only two unrelated boys. Not related to nearby FRAXA gene (Gedeon et al., 1995). May involve genes speculated to play a role in speech.

BEHAVIORAL CONSEQUENCES OF DUPLICATION SYNDROMES

Duplication syndromes describe a situation where there is evidence of a partial duplication of some genetic material. Unlike a trisomy syndrome, only a small portion of a chromosome is duplicated. The following are some selected examples of the behavioral implications of duplication syndromes.

Partial 7q Duplication

This syndrome has been reported in 37 individuals all with severe mental retardation. Findings include macrocephaly, frontal bossing, large anterior fontanelle, and hydrocephaly.

Partial Trisomy 15

There are no physical malformations associated with this rare condition, but moderate mental retardation and hyperactivity are consistent findings.

Cat Eye Syndrome (CES) (inv dup 22pter-22q11.2)

The associated phenotype of CES is highly variable but include coloboma, dysmorphology, heart defects, and mild to moderate mental retardation (McDermid, et al., 1986).

IMPLICATIONS OF CHROMOSOMAL ABNORMALITIES ON BEHAVIOR

It is hard to over emphasize the importance of the chromosomal abnormalities in our overall understanding of abnormal human behavior. Over 40 percent of severe mental retardation and 20 percent of mild mental retardation can be shown to be due to chromosomal defects at our current ability to screen. Not only do these disorders provide unambiguous cause and effect relationships between abnormal genetic endowment and abnormal behavior, they also account for a generous fraction of the etiology of mental disorders, especially mental retardation.

Chromosome abnormalities are poorly tolerated by the developing fetus. Four percent of all conceptions have chromosome abnormalities and the vast majority of these are nonviable leading to spontaneous abortion. There does not appear to be a relationship between individual chromosomes and specific types of behavior, but there is a relationship between the level of chromosomal abnormality and cognitive disruption. By the time a chromosomal problem is cytologically visible it almost always results in some degree of cognitive impairment if not outright mental retardation. Why should this be? It appears that by the time a lesion is large enough to be seen cytologically, it probably encompasses a number of genes. The cumulative loss of a number of genes appears to cause disturbances in the developing CNS resulting in mental retardation. Current estimates state that there are 50,000 to 100,000 genes in humans. There are 23 chromosome pairs. Ignoring size differences for the moment that leaves about 3,260 genes per average chromosome. If each chromosome is divided into a theoretical equal long and short arm, then that would leave about 1,610 genes per arm. Even the loss of 10 percent of a chromosome arm by deletion would effect approximately 160 genes. This argument is admittedly simplistic, but it does reinforce the point that deletions effect numerous genes, and damaging numerous genes leads to markedly abnormal behavior as represented by mental retardation.

Technology is being developed to find ever smaller deletions with the use of molecular probes that reveal small segments of missing DNA. These molecular probes are increasing the level of detection of deletions far in excess of current cytogenetic capabilities. Increasing interest is also being paid to the ends of chromosomes, the so-called *telomere*, which seem to be prone to deletion and translocation and may be another major source of mental retardation syndromes (Ledbetter, 1995). Flint, et al. (1995), for instance, tested 99 patients with idiopathic mental retardation and found that up to 6 percent of all mental retardation may be due to cryptic subtelomeric deletions.

Chapter 4

The Behavioral Consequences of Specific Single Gene Disorders

The genetic material is subject to mutation and disruption at every level of organization. We have seen the consequences of the more obvious defects in chromosome structure, and now we will review the behavioral consequences of abnormalities at a more basic level of genetic information - the gene. What follows are a number of selected *single gene* disorders with clear and well documented behavioral consequences. They are presented in order of chromosomal location.

HUNTINGTON DISEASE [4q]

Huntington disease (HD) is the most common disease of the basal ganglia. Pathology includes atrophy of the corpus striatum and depletion of interneurons in the striatum. The genetic nature of Huntington disease has been suspected since the late 1800's. The transmission is autosomal dominant. Prevalence is 40-60 cases per 1,000,000 individuals. Clinically, the condition is associated with a wide range of initial behavioral symptoms ranging from personality changes to mild movement disorder to frank psychosis and dementia. In 1983, the condition was localized to the long arm of chromosome 4, and in 1993 the gene was isolated. Molecular studies show that HD is due to the expansion of an unstable trinucleotide repeat (CAG) in a gene known as IT-15. HD is the model of a late onset genetic disorder, and also exhibits genetic anticipation (Ranen, et al., 1995). Ten percent of at risk individuals are affected by age 20, 50 percent by age 40, and nearly 100 percent by age 80 (if still alive). This late onset has traditionally made counseling very difficult. However, with the new molecular markers it is possible to determine affected status long before symptoms begin leading to a number of clinical and ethical problems. Although rare, HD is a very important disease in our understanding of the genetic basis of abnormal behavior and has been the center of intense research interest for decades.

ACUTE INTERMITTENT PORPHYRIA [11]

Acute intermittent porphyria is a disorder of porphyrin metabolism that is late onset and variable in presentation. Affected individuals often have abdominal pain, skin lesions, and peripheral neuritis. It is inherited as an autosomal dominant condition. Incidence is about 1 in 13,000 people with strong ethnic predilection. During an attack of porphyria there is excess urinary excretion of porphobilinogen (PBG). Intermittent porphyria is due to a deficiency of PBG deaminase located on chromosome 11. Behavioral aspects include abdominal pain associated with hysteria and other personality abnormalities.

PHENYLKETONURIA (PKU) [12p]

PKU is the classic disorder of amino acid metabolism. First described by Fölling in 1934, it is transmitted as an autosomal recessive and occurs in one in 10,000 births. It is caused by a disruption in the phenylalanine hydroxylase gene (PAH), which leads to an accumulation of phenylalanine in body tissues. Behaviorally, all untreated children are severely mentally retarded with less than 4 percent with IQ's above 60. Other manifestations include hyperkinetic and autistic behavior. Women who have PKU and are not on a low phenylalanine diet have children who are always mentally retarded (regardless of whether or not they have PKU). This is one of the clearest examples of a purely environmental cause of mental retardation. Evidently the mothers function with very high phenylalanine levels, but the fetal brain simply cannot develop normally in the presence of the high circulating phenylalanine. Because dietary modification can prevent the mental retardation if started early and followed strictly, most states have mandatory PKU screening for newborns. A vast amount of molecular work has gone into the study of the PAH gene. Over 200 separate mutations of the PAH gene have been detected. Furthermore, there is a correlation between some mutations and severity of phenotype prompting some investigators to suggest molecular analysis of all infants with hyperphenylalaninemia (Zschocke et al., 1995). PKU has long been a central area of research in behavioral genetics.

DENTATORUBRAL-PALLIDOLUYSIAN ATROPHY (DRPLA) [12p]

This disorder of cerebral degeneration is transmitted in an autosomal dominant manner. With a prevalence of less than one in a hundred thousand, it is quite rare. However, the onset of symptoms has a variable age of expression with a mean age of 32 years. As such, it can be added to the list of genetically determined dementia syndromes. The disorder is due to a triplet repeat (CAG) in the DRPLA gene on the short arm of chromosome 12. One interesting finding in this condition is the presence of somatic mosaicism in tissues from different parts of the brain (Onodera, et al., 1995).

WILSON DISEASE [13q14.3]

Wilson disease (WD) is a rare disorder (1/30,000) of copper metabolism inherited as a Mendelian recessive. It was first described in 1912 by Wilson as progressive hepatolenticular degeneration. The clinical symptoms are due to excessive deposition of copper in various tissues due to the abnormal transport of copper. The abnormality in copper transport is associated with an absence or very low levels of the plasma protein ceruloplasmin, which is essential for copper transportation. However, it is not defects in the ceruloplasmin gene (3q25) itself that causes WD. Behavioral manifestations of WD are due to deposition of copper into the brain disrupting normal brain function. Classically, the onset of WD is in the teenage years with behavioral changes and liver problems. If left untreated, progressive neurologic symptoms will continue. Psychiatric symptoms include hysteria, mood changes, and psychosis with dementia. If untreated, WD eventually leads to death. Wilson disease is one of the growing number of treatable genetic illnesses. D-penicillamine or zinc salts acts as copper chelators and can be preventive if initiated before permanent damage is done. Recent molecular studies have shown that WD is due to abnormalities in the gene ATB7B or WND, which is highly analogous to the Menkes disease gene and is a copper ATPase. Twenty five known mutations exist, which include 13 small deletions and insertions, seven missense mutations, 2 nonsense mutations, and three splice-site mutations. Some clinical heterogeneity, such as age of onset, appears due to differences in allelic mutations (Cox, 1995). WD has been an area of great interest in psychiatric genetic research, not only as a cause of mental illness, but because of its treatable nature.

TAY-SACHS DISEASE [15]

This disorder of sphingolipid metabolism is due to abnormal or absent hexosaminidase-A function. The result is an accumulation of gangliosides in the central nervous system leading to progressive neurological deterioration and death by age five. It is transmitted as an autosomal recessive. Behavioral manifestations include severe mental retardation and stereotypy. Well known for being over represented in the Ashkenazic Jews (1/3600) who have a 1 in 30 carrier status, which is 10 times the general population.

NEUROFIBROMATOSIS TYPE 1 [17]

This multisystem disorder effects about 1 in every 3,000 people. It is characterized by the presence of café-au-lait macules, neurofibromas, axillary and inguinal freckling, optic glioma, Lisch nodules, and osseous lesions. The gene responsible for this disorder has been localized to the pericentric region of chromosome 17 and has been cloned. Heterogeneity does not appear to be present. The basic defect appears to involve the growth and differentiation of neural crest derived tissue. A number of studies of the behavioral sequela of NF-1 have produced evidence that individuals with the disorder have an increased risk of cognitive impairment. This cognitive impairment results in a decreased average IQ for the NF-1 cohort. Some studies suggest up to 30 percent of affected individuals are mentally retarded. The cognitive impairment seems to be related to well documented visual-perceptual disturbances that are possibly due to mass effects in the CNS from fibromas. Pathologic examination reveals evidence for cortical dysgenesis.

SJÖGREN-LARSSON SYNDROME (SLS) [17p]

This is a rare autosomal recessive disorder with congenital ichthyosis, mental retardation, short stature, speech defects and spasticity (Rogers, et al., 1995). The disorder is due to the severe deficiency of the enzyme fatty aldehyde dehydrogenase (FALDH).

DISINHIBITION-DEMENTIA-PARKINSONISM-ATROPY [17q21-22]

Reported in one family by Wilhelmsen, et al. (1994), this diverse disorder includes depression, schizophrenia, impulsiveness, and neurological problems in a three generation family. These broadly defined phenotypes seem to segregate with markers on 17q. Clearly this is a preliminary finding, but if confirmed would strongly support the theory of phenotypic heterogeneity in behavior. Two of the 13 affected members had received a diagnosis of schizophrenia, suggesting that DDPA may be another rare cause of schizophrenia.

MYOTONIC MUSCULAR DYSTROPHY [MMD] [19q13]

This condition was first described in 1909 by Steinert. Clinically there is a pattern of muscle weakness, myotonia, and wasting. MMD is an autosomal dominant disorder and has been localized to chromosome 19. Behavioral manifestations often include difficult personality style with suspicious, uncooperative behavior. Hypersomnia and hypersexuality have also been noted. This condition was recently shown to be due to a trinucleotide repeat expansion, which explains the clinical evidence for "anticipation" seen in the disorder.

METACHROMATIC LEUKODYSTROPHY (SULFATIDE LIPIDOSIS) [22q]

Metachromatic leukodystrophy (infantile form) (MLD) is an

autosomal recessive disorder caused by absence of the enzyme arylsulphatase-A. Prevalence is about 1 in 40,000. Sulfatide accumulates in the central nervous system leading to demyelination. Clinically there is progressive ataxia, incoordination, and hypotonia. Behavioral manifestations include severe mental retardation. Death usually occurs by five years of age.

X-LINKED MENTAL RETARDATION SYNDROMES

Numerous X-linked syndromes of mental retardation with other behavioral abnormalities have been detected in recent years. At least 20% of all mental retardation in males and perhaps 10% of mental retardation in females can be traced to X-linked disorders. There are a least 40 described disorders of X-linked mental retardation known at this time. Male hemizygosity is the reason why more males seem to be affected with X-linked disorders. The following are selected X-linked mental retardation syndromes.

Oculocerebrorenal Syndrome [X]

This condition is a rare X-linked disorder with mental retardation as the behavioral consequence. OCRL is also known for the tendency of affected individuals to have stubbornness and tantrums. OCRL is due to a defect in OCRL-1 gene, which is involved in operation of the Golgi apparatus.

Coffin-Lowry Syndrome [X]

X-linked inheritance is strongly suggested by the presence of severely affected males and moderately affected or unaffected females. First described in the late 1960's, the syndrome is associated with growth deficiency, short stature, coarse facies, down slanting palpebral fissures, and tapering fingers. Behaviorally, there is usually moderate to severe mental retardation in males with mild retardation in obligate carrier females. Speech rarely develops. The condition is not progressive.

FG Syndrome [X]

This syndrome of imperforate anus, hypotonia, prominent forehead, and megalencephaly was first described in the 1970's. Over 30 cases have been described. Limited to males, obligate female carriers are unaffected. Behavior is typified by mental retardation. In addition, it has also been noted that these individuals are hyperactive with short attention spans leading to tantrum like episodes.

Ornithine Transcarbamylase [X]

This is an X-linked disorder where affected males have hyperammonemic coma. Heterozygote females also can experience delirium due to hyperammonemic coma.

Lesch-Nyan [X]

This condition is an X-linked recessive disorder. It was first described in 1964. This syndrome is due to a deficiency in the enzyme hypoxanthine guanine phosphoribosyl transferase (HGPRT). The disorder leads to self mutilating behavior, severe mental retardation, and early death.

Menkes Disease [X]

This condition is a disease of copper transport. Copper enters intestinal cells from dietary sources, but is not further absorbed leading to severe copper deficiency in many essential tissues. This, in turn, leads to neurological deterioration and death in early infancy. Treatment with copper histidine has been helpful. Behavioral manifestations include mental retardation. The causative gene is known as ATP7A or MNK gene. This gene is a copper-transporting ATPase. Deletions in this gene have been found in 15-20% of cases. A milder form known as *cutis laxa* is also described with no known behavioral effects.

Fragile X Syndrome [Xq27.3]

Fragile X syndrome is the most common single gene cause of mental retardation and is the second most common genetic form of mental retardation after trisomy 21. The majority of affected males have IQ's in the 20 to 50 range. It effects about 1 in a 1000 members of the general population. First discovered over 20 years ago, the descriptive title given to the syndrome is due to a detectable abnormality of X-chromosome morphology that is caused when karyotypes are prepared in a special media. The tip of the X chromosome appears fragile and to be breaking away from the body of the chromosome.

Since fragile X is an X chromosome abnormality, its effects are more pronounced in males who possess only the fragile X chromosome and a Y chromosome. Females are almost always heterozygous. Males are usually mentally retarded, but they also have a number of other behavioral problems especially learning disabilities. Female carriers are not as noticeably affected as males, but up to one third are mentally retarded or have other behavioral problems including learning disabilities.

Studies of dysmorphology in fragile X syndrome reveal high birth weight, greater than average occipital circumferences, protruding ears, hyperextensibility, and long facies. Males have been noted to have increased testicular volume. Studies in behavior show some distinctive abnormalities including hand flapping, biting, hyperactivity, and poor eye contact. A number of individuals with fragile X syndrome also meet the diagnostic criteria for autism.

Few genetic behavior disorders have enjoyed the rate of progress in the understanding of their pathogenesis than fragile X syndrome. It was specifically localized to Xq27.3 in the 1980's. In 1991, the elucidation of the exact molecular nature of the fragile X defect excited the scientific community. The gene responsible for fragile X is known as "FMR-1" and the molecular defect associated with the cytogenetic findings is a highly variable repeat of the triplet CGG. However, non repeat mutations such as deletions of sections of the FMR-1 gene also can lead to the phenotype (Wöhrle et al). The expansion of the triplet repeat is correlated with disease severity. Less than 40 repeats is considered normal, although great variability exists in exact number of repeats. From about 40 to 200 repeats is called a "premutation" because the offspring of these individuals (who are themselves symptom free) are at high risk of further expansion and having the clinical phenotype due to methylation of the DNA and lack of RNA production (Sherman, 1995). Interestingly, studies have shown that the expansion is greater when sons inherit the unstable repeat from their mothers (1.45kb on average) than when daughters inherit the unstable repeat (.76kb on average). This strongly implies that the expansion is a postzygotic event! It is further speculated that the presence of a homologous X chromosome in the female is the stabilizing factor (Loesch, et al., 1995). What the transmission of fragile X syndrome implies for behavioral genetics is that the days of simple Mendelian ratios are gone for many disorders. In its place, there will be a complicated series of risk calculations involving parent of origin effects and likelihood of triplet repeat and contraction (Ashley & Sherman, 1995).

X-LINKED HYDROCEPHALUS [Xq28]

This condition is the most common genetic form of hydrocephalus accounting for 25 percent of cases. Symptoms include hydrocephalus, clasped thumbs, and spastic paraparesis. Linkage analysis places the X-linked hydrocephalus gene at Xq28. Behavioral manifestations include moderate mental retardation. Jouet, et al. (1995) have implicated the neural cell-adhesion molecule L1 as the gene involved in not only x-linked hydrocephalus, but also MASA syndrome (mental retardation, aphasia, shuffling gait, and adducted thumbs) and spastic paraplegia type 1. The L1 molecule is known to play a large role in neuronal migration and cell adhesion suggesting that the clinical phenotypes are due to abnormalities of CNS development.

MUCOPOLYSACCHARIDOSIS TYPE II (HUNTER SYNDROME) [Xq28]

Hunter syndrome is due to a defect in the iduronate-2-sulfate-2-hydrolase enzyme, which is located at chromosome Xq28. Enzyme deficiency leads to toxic levels of heparan sulfate and dermatan sulfate, which leads to skeletal deformities, hepatosplenomegaly, and cardiopulmonary failure during childhood. Behavioral manifestations include severe hyperactivity and developmental delay leading to progressive mental retardation and death. Milder forms of Hunter syndrome are probably due to allelic variation in the causative mutations allowing some individuals to live into adulthood usually without mental retardation. Reproductive fitness is impaired to the extent that virtually all new cases are felt to be due to new mutations (Jonsson, 1995).

ANDROGEN INSENSITIVITY SYNDROME [Y]

These individuals are 46,XY in genotype, but a molecular defect causes their cells to be insensitive to the effects of androgen. They develop as normal females (although sterile with cryptorchidism) and are usually heterosexual in orientation. This condition is rare, but is of great importance in understanding the genetics of gender determination. The disorder is due to mutations in the sex determining gene (SRY) on the Y chromosome (one of only a few genes on the Y chromosome). Most cases are new mutations, but in some cases the father is mosaic 46, XY/46XY* but is fertile which suggests a post zygotic somatic mutation. In addition, SRY causes the expression of Müllerian inhibiting substance (MIS) which inhibits female development. Defects in MIS may also play a role in androgen insensitivity (Haqq, et al., 1994).

SINGLE GENE DISORDERS WITH BEHAVIORAL IMPLICATIONS THAT ARE NOT YET LOCALIZED

Tuberous Sclerosis (TS)

This condition was first reported by von Recklinghausen in 1862. Clinically, individuals with TS have seizures, hamartomas, ungual fibromas, adenoma sebaceum and hypomelanotic macules although there is marked variability of expression. Behavioral manifestations include mental retardation in 50 percent of affected individuals. Because the gene or genes for TS are not uncommon, this single disorder accounts for about 0.7 percent of all mental retardation. Transmission is considered autosomal dominant with one third of cases apparently being due to fresh mutations. There is some evidence suggesting that there is a TS gene on chromosome 9, while other evidence suggests the disorder is heterogeneous.

Olivopontocerebellar Degeneration [unknown]

This condition is typified by late onset dominantly inherited cerebellar ataxia with progressive dementing process. The mode of transmission appears to be dominant, however, OPCD probably is several specific genetic disorders with a common pathology. (Rosenburg)

Hallervorden-Spatz [unknown]

Hallervorden-Spatz disease is a rare, fatal neurodegenerative disorder with an apparent genetic etiology, although the exact mode of inheritance is unknown. Autopsy shows a rust-brown discoloration of the globus pallidus and the zona reticularis of the substantia nigra. This discoloration is apparently due to an increase in iron and other pigments. Diagnosis is mostly dependent upon clinical history, although magnetic resonance imaging may start to play a larger role. Clinical course includes neurologic symptoms (progressive dysarthria, parkinsonian tremor, and choreoathetosis) and psychiatric symptoms (depression, rage, irritability, and dementia). The age of onset is from childhood to adult, and there is an average 8 to 10 year progressive downhill course. Some evidence suggests the disorder in due to an autosomal recessive gene. However, it is an extremely rare disorder, and there is little evidence of consanguinity in the families reported, which is the opposite of what would be expected in a rare recessive trait. In addition, most cases are sporadic, even in families with a number of offspring. One alternative explanation is that the disorder is due to a new dominantly transmitted mutation. This suggests that the few families with more than one affected member may be due to parental germ line mosaicism.

Niemann-Pick [unknown]

Niemann-Pick syndrome describes several specific disorders

of sphingomyelin deposition. Clinically, there is failure to thrive and hepatomegaly. The disorder results in severe mental retardation and death in 80 percent of cases by age 4.

Hurler Syndrome [unknown]

Hurler syndrome (mucopolysaccharidosis 1-H) involves coarse facial features, deformed skull, small stature, hepatosplenomegaly, and corneal opacities. The disorder is progressive with death usually by age 14. Behavioral manifestations include progressive mental retardation. This condition is due to a defect in alpha-L-iduronidase and is inherited as an autosomal recessive trait.

Sanfillipo (Mucopolysaccharidosis III) [unknown]

This condition is characterized clinically by mild dysmorphology with increasing behavioral disturbance leading to death usually be age twenty. Problems include hyperactivity, impulsiveness, and progressive retardation. It is caused by a deficiency in the enzyme N-ac-glucosamine-6-sulphate sulphatase. Treatment includes the use of neuroleptics and sedatives.

Rett Syndrome [unknown]

Rett syndrome involves progressive neurologic deterioration and encephalopathy with onset in infancy. It occurs in about 1 in 12,000 females. Mental retardation is the norm. Communication disorders are very prevalent, as is stereotypy. Other abnormalities include seizures and microcephaly. The genetic location is mystery. It was long speculated to be an X-linked disorder due to the fact that no males have been found to have the disease and that there was a very high MZ concordance in affected female twins. However, recent evidence involving lack of isodisomy in twins and normal X-inactivation patterns suggests that it is not an X-linked condition.

Galactosemia [unknown]

Galactosemia is the most common genetic disorder of

carbohydrate metabolism. Galactosemia results from lack of the enzyme galactose-1-phosphate uridyl transferase, which converts galactose to glucose. The disease appears in infancy and if untreated leads to mental retardation, lethargy, liver failure, and cataracts. Elimination of lactose from the diet early in infancy can prevent many symptoms from developing. Transmission is as an autosomal recessive.

Amyotrophic Lateral Sclerosis

Amyotrophic lateral sclerosis was first defined by J.M. Charcot in 1869 and is a degenerative disease of the motor neurons. Behaviorally, it can result in slowed speech, (spastic dysarthria), and emotional lability due to corticobulbar involvement. As much as 10 percent of cases of ALS are inherited as a dominant trait and many sporadic cases may have a genetic/molecular basis.

Smith-Lemli-Opitz Syndrome

This multiple congenital anomaly syndrome with associated severe mental retardation and failure to thrive is an autosomal recessive trait. The incidence is 1 in 30,000 births. Alley, et al. (1995) have presented evidence that SLOS in located at 7q32.1 based on a patient with a translocation.

IMPLICATIONS OF THE BEHAVIORAL CONSEQUENCES OF SINGLE GENE DISORDERS

The causal relation between single gene defects and behavioral abnormality is unequivocal. The numerous, well documented examples presented here are but a fraction of the single gene disorders with known behavioral consequences. One could even go as far as to speculate that if every gene were isolated and all their defects noted that the vast majority of all mental illness would be shown to be due to specific single gene disorders. However, at this time, only a fraction of mental illness can be attributed to specific single gene disorders, but, likewise, only a small fraction of genes have been discovered. One can imagine a time when disorders such as schizophrenia will be described in terms of a number of specific single gene defects.

At least two mechanisms exist to explain how single gene defects lead to abnormal behavior. Some defects appear to distort the normal development of the brain. Other defects apparently lead to accumulating toxic substances, which then damage the normally formed brain. Another important aspect of the genetics of single gene disorders is the variable expression of the behavioral components. In many conditions, only a percentage of affected individuals have abnormal behavior, and of those affected, there is a wide range of severity. The implication is that the relationship between the genotype and phenotype is not tight in the genetics of behavioral disorders. Factors such as epistasis, allelic heterogeneity, and randomness in development help to create variation in the expression of these traits.

Chapter 5

Approaches to Studying the Genetics of Broad Behavioral Phenotypes

Chromosome disorders and single gene disorders are a major source of abnormal human behavior. Ultimately, they will likely account for the bulk of abnormal human behavior. Until that time, however, there have been concerted attempts to study the genetics of abnormal behavior by studying the genetics of the broad behavioral phenotypes. In fact, the bulk of human behavioral genetic research has been on the study of broad behavioral phenotypes. This is unfortunate, because these approaches seem increasingly unlikely to yield the detailed genetic answers that are being sought. None of them, for instance, are designed to localize a gene involved in behavior or to even clearly specify the exact genetic transmission. At best, these strategies can confirm the general genetic basis of a disorder and point to a direction for more detailed study. The traditional research strategies used to study broad behavioral phenotypes are as follows: family studies, twin studies, adoption studies, high risk studies, and segregation analysis. Each of these will be discussed and their strengths and limitations will be analyzed.

THE ROLE OF FAMILY STUDIES IN STUDYING BROAD BEHAVIORAL PHENOTYPES

Family studies are a logical first step in helping to determine if there is a genetic contribution to a disease. The simplest approach is to identify individuals with the disorder and then study their extended families to determine if other members also have the disorder. Comparisons can then be made as to the number of parents, siblings, and offspring with the disorder versus the number of unaffected persons. These percentages can then be compared to levels of the illness in the general population and a relative risk can be determined. Early investigators often looked at this data in the form of pedigrees and found patterns suggestive of a Mendelian type of transmission. For example, Huntington disease was noted to be a Mendelian dominant trait based on the observation that around 50 percent of offspring were affected and 50 percent of siblings were affected. However, while single families might appear to have a Mendelian transmission, the broad behavioral phenotypes have shown no consistent pattern of transmission. Early attempts to show that all mental retardation or all schizophrenia was due to a single Mendelian gene are now seen as embarrassingly simplistic.

Requirements for Valid Family Studies

Another problem with family studies is that they are highly prone to bias due to poor design. Very specific methods must be used in order to have any confidence in the results. Many early family studies were severely flawed because of lack of attention to some aspect of study design. A valid family study requires attention to all of the following details.

Cases should be identified by one member of a family with the disorder in question. This individual is then identified as the *proband* or *index case*. The proband must be identified in a non-biased manner. The best way to do this is to define a population such as a clinic or hospital and take the next "x" number of patients with the disorder. This would limit the bias to the inherent bias of the different settings, a bias that can be accounted for. The next step is to develop reasonable *exclusion criteria,* which is the rules for determining which patients can and cannot be included in the study. Often patients cannot be included if they have dual diagnoses, which can confound the study design or if they live too far away from the study center to be followed up. The question that must be asked in developing exclusion criteria is: Does the excluded population differ from the included population in any meaningful way? Efforts must be made to be certain that they are similar. Many of the early family studies in psychiatric genetics were flawed due to selection bias. For example, some studies used only families with several affected members, thus over representing the familial nature of the disorder.

The next major problem area is diagnosis and data collection. Family members must be interviewed in person. Many early studies relied upon family reports or minimal clinic records to determine a person's *affection status* (whether they have the disorder in question). This type of data collection called the *family history method* and it is particularly unreliable. The *family study method*, which involves direct interviews with all included family members, is essential to proper study design.

A further design requirement is that diagnoses must be made with consistent, accepted diagnostic criteria. This design requirement is very difficult in broad behavioral phenotypes since diagnostic criteria have typically been vague and have shifted significantly over the years. This makes comparisons between early family studies and modern ones very difficult. Even if long term diagnostic consistency is unreliable, it is even more critical that there is diagnostic consistency within the group (reliability). In addition, diagnoses should be as blind as possible both to the patients and to their position in the overall family. Many early studies were not blind, thus introducing tremendous bias.

Finally, comparisons to normal control populations must be from a group derived from the same population and diagnosed using the same criteria. Any deviation from these standards can result in significant biases that invalidate the study. Several early studies did not use control groups and relied on population data as a control, their positive findings later turned out to be due entirely to different diagnostic criteria between the study group (more inclusive) and the control group (less inclusive).

Risk Assessment

Ultimately, all family study designs determine *relative risk*. As an example, disease Z is present in 15 percent of siblings of the probands while only 1 percent of the siblings of the control group are affected. On face value, this appears to be a relative risk of 15/1 or 15. In other words, siblings of affected persons have a 15 times greater risk of the disorder than siblings of an unaffected person. However, these risks must be adjusted. It was noted early on that many persons studied had not entered the *age of risk,* which is the minimum age when the condition is diagnosable. Some conditions are apparent at birth, like cleft palate, so there is no adjustment, but other conditions only manifest late in life, so younger persons have more risk of the disease than an older person. Two methods have been developed to determine these *lifetime morbid risks.* The most common technique is the *Weinberg abridged method.* Another technique is *survival analysis,* which bases age adjustments on the natural history of other family members. All methods are dependent upon empiric lifetime risk data, and cannot be done without it. Failure to adjust for age of risk will reduce the apparent risk of family members, especially in late onset disorders, and minimize the familial nature of the condition. When these procedures are carried out appropriately, a *relative risk* calculation can be done.

The Limitations and Implications of the Family Study Approach

Several limitations severely hamper the usefulness of family studies. The first is that even a positive family study with a high relative risk does not necessarily indicate that the risk is due to genetic factors. An absurd example that is often used to reinforce this point is that the language a family speaks is highly familial, but obviously not genetic. On the other hand, lack of familiality speaks against heredity. Family studies tend to work best with rare disorders that have high penetrance. Huntington disease is a good example of this. Common disorders like broad behavioral phenotypes have less satisfactory results. Family studies should not be confused with epidemiologic studies, which are more concerned with incidence and prevalence. However, most family studies generate epidemiologic data as a byproduct, and conversely good family studies need population based prevalence and age of onset data to be analyzed. Finally, one byproduct of family studies is the use of relative risk data in determining empirical risk for disease transmission in genetic counseling. In schizophrenia, for instance, there is about a 10 percent relative risk to offspring of a single affected parent (although this number varies by study). This number is often given to prospective parents (when one is affected with schizophrenia) to suggest that they will have a one in ten chance of a child with schizophrenia. The problem with this approach is that empiric risk is quite different from the genetic risk of a disease with known mode of transmission. In empiric risk, the overall risk of a large group may be 10 percent, but individuals may have 50 percent, 25 percent, zero percent, or some other level of risk that is hidden in the overall group numbers.

THE ROLE OF TWIN STUDIES IN STUDYING BROAD BEHAVIORAL PHENOTYPES

The use of the natural phenomenon of twinning as a strategy to study the "nature versus nurture" question was first suggested by Francis Galton in 1875. Since its first systematic application to a psychiatric illness (schizophrenia) in 1928, it has been an invaluable tool to help decipher the question of whether inheritance plays a role in broad behavioral phenotypes. Twin studies go much further than family studies in addressing the question of whether genetic factors are involved in the transmission of the disorder, however, they too have significant limitations.

Design and Interpretation of Twin Studies

Twins can be either identical *(monozygotic-MZ)*, which means that they developed from the same fertilized egg cell (zygote) or fraternal *(dizygotic-DZ)*, which indicates that they developed from two separately fertilized eggs. The monozygotic (MZ) twins share identical genotypes at conception, while the dizygotic (DZ) twins share only 50 percent of the same genes just as any siblings would. Twin births vary in incidence according to a number of factors, but the average rate is about 1 per 100 births. DZ births are more common that MZ births in an approximately 2 to 1 ratio. Since nature (genetics) and nurture (environment) are both operative in development, a long standing question has been the relative influence of each. MZ twins seemed to be a mechanism to solve this question, the assumption being that identical twins should be identical for a trait if it is genetic, whereas DZ twins should only be faced with a maximum of 50 percent risk. When both twins are similar for a trait or illness they are called *concordant*, if dissimilar they are referred to as *discordant*.

There are two major ways to determine concordance in psychiatric genetic research. In *pair-wise concordance*, each twin pair is simply described as concordant or discordant regardless of ascertainment. In *proband-wise concordance*, each ill twin is identified as a proband so that concordant pairs are counted twice. This method is more appropriate for analysis of twin samples collected without reference to the studied illness, however, most psychiatric illnesses are rare enough that twin samples collected without reference to the illness are not practical, so it is rarely used.

MZ twins, either concordant or discordant for a disorder, were reported as single case studies for many years, but such anecdotal reports are subject to bias. Early on, researchers realized that large numbers of twins would have to be studied to make any valid comparisons. In addition, the way the twins were collected for the study also had high potential for bias, so care must be taken to collect a series without regard to zygosity or concordance. When these criteria are met, the comparison of concordance rates for MZ and DZ twins can produce an approximation of *heritability*. Heritability is a measure of the genetic influence of the phenotype. As a percentage, it is calculated as the variance due to genetic influences over the total variance of the phenotype. Holzinger (1929) quantified this for twin studies as:

$$H = \text{Heritability} = \frac{(\%\text{ MZ concordance} - \%\text{ DZ concordance})}{(100 - \%\text{ DZ concordance})}$$

This index is meant to convey the concept of heritability, and is not meant to be seen as a precise measure. However, when measuring something like ABO types the MZ concordance is 100 percent, which gives an H = 1. Whereas in something truly environmental, like infection, the MZ concordance is roughly equal to DZ concordance and H = 0.

Another use of twin studies is to give clues to the mode of inheritance. For a single gene trait, the ratio of MZ to DZ twins should be two for dominant traits and four for recessive traits assuming the parents are heterozygous. If the ratio is less than two, traditionally, it was felt to indicate significant environmental factors. A ratio greater than around four has been explained by polygenic inheritance.

Misconceptions and Limitations of Twin Studies

Twin studies have a rich and productive history in psychiatric genetic research. However, they also have a long history of controversy. The source of this controversy can be divided into two schools. One school of thought insists that twin studies overestimate the role of genetics, the other that twin studies underestimate the role of inheritance. The major basis for the belief that twin studies overestimate the role of genetics is the view that the process of being a twin somehow explains the increased MZ concordance. This view is mostly promulgated by practitioners of psychodynamic based theories that are antagonistic to genetic determinism. As such, some of the criticisms have a somewhat desperate, even obscurest, quality to them. For instance, a major criticism has been that MZ twins are different then DZ twins and that this theoretical difference leads to alterations in MZ and DZ twin concordances. The first argument is that the process of being a MZ twin is so debilitating to the psyche that just being a MZ twin increases the risk for psychiatric illness. The problem with this theory is that there is no evidence whatsoever that being a MZ twin increases the risk of any illness. In addition, the VA twin study demonstrated that twins have the same incidence of psychiatric disease as non-twins. This criticism must, therefore, be seen with great skepticism.

A second critique is that DZ twins are exposed to different environments while MZ twins are treated alike. In other words, DZ and MZ twins have different environments and the difference in concordance is simply due to the different environments. The problem with this criticism is that there is no plausible, consistent environmental difference that can be measured to explain this difference. Surely, a variable with the ability to alter concordance rates would be robust enough to be easily measured. No such evidence exists. In fact, no evidence exists to show that treating children alike increases the level of any psychopathology. Even without evidence of this theoretical environmental factor, however, many found the suggestion quite satisfying and claimed it invalidated twin studies. In an attempt to answer this criticism, various investigators decided to look at MZ twins who had been raised apart in the belief that separate environments would nullify the shared environment argument. These studies are quite difficult, because of the rarity of MZ twins reared apart, but an increasing data base has been compiled that shows agreement with the DZ and MZ concordance rates of earlier studies. In addition, Kendler (1983) reviewed the question of shared environment as a cause of MZ-DZ concordance differences and found no evidence to support it, which leads to the conclusion that the shared environment argument is invalid.

The second school of thought is that twin studies are flawed because they are biased to under represent the role of genetics, especially in complex behavioral disorders. Perhaps the biggest misconception in twin studies is the idea that MZ twins are biologically "identical." This concept derives from the fact that identical twins have the same initial genetic endowment, but from that instant on they begin to diverge biologically. Somatic mutation and chromosomal problems early in development can lead to variation in MZ twins. Even without mutation, development ensures that significant variation occurs as random events are magnified throughout development. The result is that by the time of birth there is significant variation in MZ twins biologic systems. The brain, as the most complicated structure, is likely to have the greatest variation. This lack of biological identicalness in MZ twins can only lead to underestimating heritability since biological differences after conception can only decrease concordance.

A closely related aspect of this same issue is the contention that all differences between MZ twins must be explained by environmental factors. This is often apparent in statements like. Disease "X" must be, in part, due to environmental factors since the MZ concordance is only 80 percent. Failure to attain 100 percent concordance can be due to many factors, most of which are not typically considered environmental. Part of this problem goes back to the original nature versus nurture dichotomy in behavior disorders, where nature has become very narrowly defined as just the genetic contribution at conception, and environment is defined as everything else. This is misleading, because the part of "environment" that has come to symbolize environment most closely is, in fact, the very narrow portion that mostly involves life events and social interactions after the child is born. This leaves a large amount of development and other biological events that clearly have great effects on final phenotypic outcome in a gray area. The solution is to call this third variable *developmental plasticity* and clearly separate it from environment and from the genetic endowment. As a result, one would have to modify the above statement. Disease "X" must be in part due to either developmental plasticity and/or environmental factors since the MZ concordance is only 80 percent.

In addition to the theoretical misconceptions noted above, there are other practical limitations of twin studies that can interfere with study interpretation. First, ascertainment is a major area of potential bias. The most blatant example of this results from beginning a study by looking for twins with a particular disorder. This would clearly skew the findings toward higher heritability. It is essential that twins used for a study be identified in some systematic way, such as taking the first 50 twins from a hospital or clinic list. In addition, even the degree of severity can bias the study because severity of illness is often correlated with higher concordance rates. Another problem area is the diagnostic criteria. For instance, if the exact diagnostic criteria for schizophrenia were known, it would be simple to determine affection status. In reality, studies vary in the strictness of diagnostic criteria with direct effects on MZ and DZ concordance. Some of the twin studies with the highest concordances have been criticized for being too inclusive, while the few studies with low concordances have been criticized for being too diagnostically strict. A third problem, particularly with the early twin studies, is the question of zygosity. Determining which twin pairs are MZ and DZ on physical appearance or fingerprints alone is insufficient. Detailed tests of zygosity must be done. Studies done prior to these exhaustive tests are, therefore, open to question.

Twin studies have played such a large role in psychiatric genetic research that there is a tendency to overemphasize their significance. They are important and are clearly helpful in documenting that a broad behavioral phenotype, like schizophrenia, has a high heritability suggesting that further research be done.

However, twin studies in broad behavioral phenotypes also have severe shortcomings. For instance, they tell us nothing about many of the questions that we really want to know, like the exact mode of transmission, the level of heterogeneity, or the location of a gene involved in the disorder. For most broad behavioral phenotypes, the twin study approach has been maximized and is of little further use.

Studies of the Offspring of Discordant Twins

A recent trend in twin study research is to study the offspring of MZ twins who are discordant for a disorder. This approach was pioneered by Fischer in 1971. If the factors that led to discordance in the MZ twin pair are not genetic, then the rate of illness in the offspring of the unaffected twin would be less than the rate in the offspring of the affected twin. A finding that offspring affection rates are the same would support the genetic hypothesis and indicate that some factor may reduce penetrance of the trait in the unaffected twin.

Using Twin Studies to Study Comorbidity

A recent novel use of twin data was reported by Neale and Kendler in 1995. This method uses MZ and DZ twins to model the type of comorbid relationship between two multifactorial disorders such as depression and generalized anxiety disorder. Such models can test various relationship scenarios such as chance, sampling bias, and reciprocal causation.

ADOPTION STUDIES

It is an historical quirk of behavioral disorders that environment in the form of life events has been emphasized out of proportion to other environmental influences and to the evidence supporting it. This overemphasis of the role of life events in the etiology of behavioral disorders resulted in repeated criticisms of family and twin studies. To help solve this impasse a novel research approach was developed – the adoption study. The adoption study is a research strategy evolved to answer a specific question. What is more important childrearing or biological/genetic influences in the familial transmission of major psychiatric disorders?

Design and Interpretation of Adoption Studies

Three general types of adoption studies have been done. The *adoptee family method* is performed by compiling an index of mentally ill individuals that have a history of being adopted. This cohort is matched with a group of non-mentally ill individuals who were also adopted. The incidence of illness in the biologic and adoptive parents of both groups are determined. If genetics plays no role in the disorder, the incidence of mental illness should be about the same in the index and control groups of biologic parents. If genetics does play a role then the incidence should be higher in the biologic parents of the index group. The *adoptive study method* is performed by comparing rates of mental illness in adopted away children of ill parents versus adopted away children of well parents. Finally, the *cross-fostering study* compares the rate of mental illness in the children of ill biologic parents that have been raised by well adopted parents with the rate of mental illness in the children of well parents that have been raised by mentally ill parents.

Limitations of Adoption Studies

Adoption studies are very difficult to complete. Problems with confidentiality make it very difficult to find the subject material. Most adoption studies have been done in Scandinavian countries where central registries of both adoptions and mental illness are kept and the rules of confidentiality are more lax. As in family and twin studies, ascertainment and diagnostic issues are very important and must be approached with care. However, there are some specific problems associated with adoption studies. An important problem is that the time of adoption often varies. Some children may be adopted out at birth, others after a few days, and others after a few months. Some have questioned whether the two months that the child stays with the parent is long enough for some cultural transmission to occur, although this question actually seems more relevant to theories of viral transmission. An additional problem is that adoption studies cannot answer the question of whether the intrauterine environment could be responsible for the disease. Strategies, such as comparing the risk of affected mothers to affected fathers, have been tried to help answer this question.

It has also been pointed out that adoption studies are based on an assumption that there is not anything particular about those parents that give up their children for adoption. It has been argued that mentally ill parents are much more likely to give children up for adoption than non-mentally ill parents. Others have argued that the adoptive parents may also be distinct and they point to the selection process by adoption agencies that select for only the best parents. However, the effect of the last two criticisms would likely only minimize the genetic contribution, making these studies even more powerful in determining genetic influences. Finally, and most importantly, adoption studies are helpful only in a qualitative sense, but they are not useful in any quantitative sense. In addition, just like twin and family studies, adoption studies really do not bring us any closer to the most important genetic questions concerning mode of transmission and location of genes.

HIGH RISK STUDIES

If a psychiatric disorder is suspected to have a genetic component, one research strategy is to study the children of parents affected with the disorder in question, the so-called *high risk study*. These children are felt to be at *high risk* for the disorder regardless of the mode of transmission. The hope is that by studying the children in depth as they develop, contributing factors can be determined. If careful study is paid to the numerous environmental influences, perhaps one might show some predictive potential. Likewise, lack of any environmental correlation suggests that the biologic influences are the most important factor.

The Design and Interpretation of High Risk Studies

High risk studies begin with the ascertainment of an accurately diagnosed parent as a proband. Then, depending on the study, the determination of non-mental illness in the other parent is made. The children of the probands are established into a cohort. This cohort is then followed prospectively with any number of measurements. The high risk studies are designed to answer several questions. First, by comparing the cohort with an age matched control group, early differences in development can be determined. Second, by comparing early development of those that ultimately become ill with those that do not, it is postulated that the earliest hints of pathology or possible triggers can be determined. High risk studies are hampered by the same problems of ascertainment, diagnosis, and logistics as the other studies. They also can be seriously hampered by assortative mating. In addition, unless 1 to 1 cause and effect environmental variables are postulated, any environmental effect is likely to be incremental and subtle. This obstacle can only be overcome with ever increasing sample sizes. The future of high risk studies would be greatly enhanced by the presence of trait dependent markers such as a molecular marker linked to the disease phenotype.

SEGREGATION ANALYSIS

With the advent of the modern high speed computer, massive amounts of pedigree data can be analyzed and reanalyzed in seconds leading to a number of computerized strategies for determining the mode of transmission. The problem with this approach for behavior disorders is that the high prevalences and relatively low genetic risk often supports several possible outcomes mostly dependent upon which initial variables are used. This often leads to contradictory studies and vague conclusions that simply end up restating what is already known - that the genetic transmission in broad behavioral disorders is complex. To illustrate this point, an investigator once studied medical students and their families and did a segregation analysis that found that medical school attendance could be explained by a single recessive gene! Furthermore, with the advent of the linkage study approach, much more direct methods are now available to study mode of transmission.

Chapter 6

The Genetics of Broad Behavioral Phenotypes

There are relatively few behavioral phenotypes in abnormal human behavior compared to the large numbers of individuals affected. To help define this point, the term *broad behavioral phenotype* is used to describe most psychiatric syndromes. For a variety of reasons, it appears extremely unlikely that the clinical homogeneity of these broad behavioral phenotypes reflects specific etiologic homogeneity. However, for many years these broad behavioral phenotypes have been investigated through the various study designs available, and the results comprise the bulk of the data base of psychiatric genetics. The following is a summary of the findings and implications of these studies.

THE GENETICS OF MENTAL RETARDATION

The broad behavioral phenotype that has met with the most success in defining genetic factors is mental retardation. It is very instructive, therefore, to begin the review of the genetics of broad behavioral phenotypes with a review of the genetics of mental retardation. The progress in mental retardation is due primarily to the fact that a number of subtypes of mental retardation have been described based on differences in dysmorphology, metabolic abnormalities, and chromosomal abnormalities.

Epidemiology of Mental Retardation

Mental retardation describes a behaviorally defined set of

disorders of intellectual and adaptive function with an overall prevalence rate of approximately 1 percent. As shown in Table One, the specific diagnosis of mental retardation is based on an IQ less than 70, impaired general adaptation, and onset before age 17 (to differentiate from dementia). Mental retardation is further subtyped as being either mild (IQ 55-70, comprising 85 percent of the mentally retarded population), moderate (IQ 40 - 55, comprising 10 percent of the mentally retarded population), severe (IQ 25-40, comprising 3-4 percent of the mentally retarded population), or profound (IQ < 25, comprising only 1-2 of the mentally retarded population). With a U.S. population of 250 million there are about 2.5 million mentally retarded individuals.

TABLE ONE:
ETIOLOGY OF MENTAL RETARDATION BY SEVERITY

DEGREE OF SEVERITY	IQ RANGE	PERCENT OF ALL MR	ETIOLOGY
MILD	55-70	85 %	Familial - Unknown
MODERATE	40-54	10 %	Congenital - Genetic
SEVERE	25-39	3-4 %	Congenital - Genetic
PROFOUND	< 25	1-2 %	Congenital - Genetic

The Causes of Mental Retardation

Until the nineteenth century, mental retardation (or idiocy as it was known as previously) was seen as a fairly homogeneous group of conditions somewhat analogous to how we currently view schizophrenia. As the result of painstaking methodical studies by countless individuals we now know that the behavioral homogeneity comprising mental retardation masks a vast etiologic heterogeneity. A large portion of persons with mild mental retardation are simply the tail end of the normal distribution. These individuals are notable for having little if any dysmorphology, no specific metabolic problems, normal mortality and morbidity, and usually a nonspecific family history of lower than average intelligence. This group may comprise as much as 85 percent of all mental retardation. These individuals are usually detected in the school setting and often function quite well in semi-structured settings. Individuals with IQ's less that 55 can often be shown to have a fairly specific cause for their mental retardation. Among the known causes are prenatal trauma in the form of toxins, anoxia, and infections, and genetic causes including a host of chromosomal and single gene disorders. The more severely an individual with mental retardation is affected the more likely the mental retardation can be shown to be due to a specific cause. This is not to say that all cases have a specific cause, because the limitations in our knowledge still allow for a large portion of cases of mental retardation (up to 50 percent) to fall under the category of unknown causation.

In terms of genetic causation, there are a number of well-described syndromes and anomalies that produce phenotypes with subnormal mental abilities. Many chromosomal abnormalities produce mental retardation in addition to physical abnormalities. In addition, there are literally hundreds of single gene disorders in which mental retardation is over represented. Many of these chromosomal and single gene causes of mental retardation have been discussed previously. Table Two, however, shows a number of less common genetically caused forms of mental retardation.

TABLE TWO:
SELECTED RARE DYSMORPHIC SYNDROMES
ASSOCIATED WITH MENTAL RETARDATION

Syndrome	(# of Cases)	Major Findings

- Acrodysostosis (>20) Short hands, peripheral dysostosis, small nose, virtually all individuals affected with mental retardation (>90%). Robinow: AJD Child 121:195, 1971.
- Albright Osteodystrophy (>100) Short metacarpals, round face, hypocalcemia, mean IQ of 60 with 70% mental retardation. Fitch: AJMG 11:11, 1982.
- Baller-Gerold (>8) Craniosyostosis, radial aplasia, 50% have mental retardation. Pelias: AJMG 10:133, 1981.
- Camptomelic Dysplasia (> 4) Bowed tibiae, growth deficiency, flat facies,

early death, severe mental retardation in the few survivors. Houston: AJMG 15:3, 1983.

- Carpenter (>100)Acrocephaly, polydactyly, syndactyly, inner canthi displacement, mental retardation over represented, IQ range 60-104. Robinson: AJMG 20:461,1985.
- Craniodigital (Scott) (> 4) Facial anomalies, finger and toe syndactyly, delayed skeletal growth, and mental retardation. Lorenz: AJMG 37: 224,1990.
- Dyggve-Melchior-Clausen (>100) Short stature, restricted joint mobility, spine problems, microcephaly, and mental retardation. Bonafede: Clin Gen 14:24,1978.
- Fanconi- Pancytopenia (>160) Radial hypoplasia, hyperpigmentation, pancytopenia, 25% have mental retardation. Autosomal recessive. Glanz: J Med Gen 19:412, 1982.
- Fraser (>90) Cryptophthalmos, auricle defects, genital anomalies, and mental retardation. Thomas: AJMG 25:85,1986.
- Lambotte (> 4) Microcephaly, hooked nose, IUGR, severe mental retardation, early death. Verloes: AJMG 37:119, 1990.
- Langer-Giedion (>30) Multiple exostosis, large nose, unusual facies, microcephaly, mild to severe mental retardation (70%). Langer: AJMG 19:81, 1984.
- Lenz-Majewski Hyperostosis (>8) Dense thick bones, symphalangism, moderate to severe mental retardation. Early death. Gorlin: Radiol 149:129,1983.
- Michels (> 5) Craniosynostosis, lid abnormalities, ptosis, borderline intellectual function, speech and language delay. Cunniff: AJMG 37:28,1990.
- Mietens (> 4) Corneal opacity, narrow alae nasi, elbow flexion contractures, average IQ 75. Mietens: J Ped 69:624,1966.
- Oral-facial- digital, Type I (>130) Oral frenula and clefts, alae nasi hypoplasia, digital asymmetry, mental retardation in over 50%, average IQ is 70, 20% have agenesis of the corpus callosum or abnormalities of the gray matter. Dominant, lethal in males. Anneren: Clin Gen 26: 178, 1984.

TABLE TWO: (CONT)

- Oto-palato- digital, Type I (>50) Deafness, cleft palate, broad digits with short nails, speech abnormalities, mental retardation with an average IQ of 80. Gorlin: Oral Surg 35:218, 1973.
- Oto-palato- digital, Type II (> 9) Ocular hypertelorism, overlapping fingers, small mouth, hearing loss, cryptorchidism, and mental retardation. Brewster: AJMG 20: 249,1985.
- Rhizomelic-Chondrodysplasis Punctata (>40) Short humeri and femora, vertebral problems, early death, microcephaly, severe mental retardation. Autosomal recessive, may be a peroxisome disorder. Schutgens: Eur J Ped 144: 430, 1986.
- Roberts-SC- Phocomelia (>20) Hypomelia, midfacial defects, growth deficiency, decreased IQ with over representation of mental retardation. Parry: AJMG 24:653, 1986.
- Ruvalcaba (>10) Hypoplastic alae nasi, small mouth, short metacarpals, mental retardation. Bianchi: Eur J Ped 142:301,1984.
- Shprintzen (>40) Small stature, hearing loss, cleft palate, prominent nose, long face, cardiac septum abnormalities, average IQ 80, increased mental retardation and learning problems. Shprintzen: Ped 67:167, 1981.
- Silver-Russell (>50) Triangular facies, short stature, cognitive impairment with 32% in mentally retarded range. Lai: Arch Dis Child 71: 490-96, 1994.

Implications for the Genetics of Abnormal Human Behavior

The broad behavioral phenotype of mental retardation is due to a number of causes, especially genetic ones. The known causes have one thing in common. They all lead to direct damage to the developing brain. This can be in the form of toxins or infection damaging a normally developing brain, or in the case of the genetic causes, a brain that is developing abnormally due to errors in the molecular information necessary to develop a normal brain. Chromosome disorders imply large scale disruptions in normal development. Single gene disorders can cause damage by inborn errors of metabolism leading to abnormal storage by defective transport mechanisms or by the loss of critical steps in development. Whatever the cause, mental retardation results when there is a gross insult to normal brain development. Each of the known genetic causes seems to explain only a small part of the large number of cases of mental retardation, but as can be seen in Table Three, when summarized they begin to account for sizable portion of all mental retardation.

TABLE THREE:
PERCENT OF ALL MENTAL RETARDATION
DUE TO CHROMOSOMAL DISORDERS (MALES ONLY)
(Prevalence of 1.0 %)

Chromosome Abnormality	Incidence	%MR	% of all MR attributable to disease phenotype
(1) Trisomy 21	1/580	99 %	17.1 %
(2) Klinefelter Syndrome	1/850	20 %	2.4 %
(3) XYY Syndrome	1/900	20 %	2.2 %
(4) Trisomy 18	1/9000	99 %	1.1 %
(5) Monosomy 5p	1/50,000	100 %	0.2 %
(6) WAGR Monosomy	1/100,000	100 %	0.1 %
(7) Beckwith Trisomy	1/100,000	50 %	0.05 %
Total			23.15 %

Just the more common forms of chromosomal abnormality account for over 23 percent of mental retardation in males. Rare disorders begin to add up also when the sheer number of such disorders is accounted for. Fragile-X and similar single gene disorders could generate an equally impressive portion of the total amount of mental retardation. Together, known genetic disorders already account for over 35 percent of all cases of mental retardation. The study of mental retardation provides the best and most convincing evidence that we have to support the argument that other broad behavioral phenotypes are masking a tremendous underlying genetic heterogeneity.

THE GENETICS OF SCHIZOPHRENIA

Schizophrenia was recognized as a distinct subset of mental illness by the pioneering work of Emil Kraepelin and Eugene Bleuler. It is of note that both of these individuals believed that hereditary factors were important in its etiology. Bleuler (1857-1939) states in his monograph *Dementia Praecox and the Group of Schizophrenias* (1911) that "heredity does play its role in the etiology of schizophrenia, but the extent and kind of its influence cannot as yet be stated." In the intervening 80 years, we have a great deal more data supporting the role of genetics in the etiology of schizophrenia, but we still cannot define the specific role of the genetic influences. When we look to find supporting evidence for the role of genetics in schizophrenia we still look to three familiar sources – family, twin, and adoption studies.

Family Studies

As is the case with most genetic disorders, the familial nature of schizophrenia was noted anecdotally from clinical observation long before it was studied scientifically. The first scientific attempts at studying this familial association were reported in 1916 by Ernst Rüdin. His monograph on the incidence of schizophrenia in the siblings of 701 identified sibships initiated the modern era of psychiatric genetics. Since Rüdin's initial studies, family studies of schizophrenia have shown a near consistent increase risk to family members over that of the general population regardless of degree of relationship (See Table Four).

Comparisons between schizophrenia family studies over the past 70 years are difficult because of the large number of different diagnostic structures used and the different ascertainments. A number of criticisms can be leveled against these studies, particularly those done up to the 1980's. Criticisms include complaints that most studies did not include a control group, that all relatives were not accounted for allowing for differential inclusion, that diagnostic criteria were not structured, that interviewing was not always done the same way, and that diagnoses were not made blind to the relationship to the proband. Many of these criticisms have been answered in the studies done over the past ten years and the results have been similar to the earlier studies.

TABLE FOUR:
RESULTS FROM SELECTED FAMILY STUDIES IN SCHIZOPHRENIA

Year	Investigator	Study Results
1916	Rüdin	First major study. 701 probands identified. Risk of schizophrenia in children 6 % if one parent affected, increasing to 23 % if both parents affected. Risk to first degree relatives 4.5 %.
1921	Hoffman	Risk to children of schizophrenic parent 9 %.
1923	Kahn	When both parents are schizophrenic risk to children is 53% (9 of 17).
1927	Schulz	Showed increased familial risk.
1928	Luxenburger	Showed increased familial risk.
1929	Brugger	Showed increased familial risk.
1932	Oppler	Risk to children of schizophrenic parent 10 %.
1933	Klemperer	Showed increased familial risk.
1933	Gengnagel	Risk to children of schizophrenic parent 8 %.
1935	Sjögren	Showed increased familial risk.
1935	Stömgren	Showed increased familial risk.
1938	Kallmann	1087 families were studied. Risk to children 15 % if one parent affected. If both parents affected then 46 % risk (16 of 35).
1940	Schulz	If both parents affected then 30 % risk to children (18 of 59).
1946	Kallmann	Spouses have a 2 % risk. Step sibs have a 2 % risk. Half sibs have a 7 % risk. Parents have a 9 % risk. Full sibs have a 14 % risk.
1951	Fremming	Showed increased familial risk.
1952	Elsässer	If both parents affected then risk for children is 27% (15 of 56).
1957	Lewis	If both parents are affected, then risk for children is 15% (4 of 27).

TABLE FOUR: (CONT)

1962	Garrone	Risk to children of schizophrenic parent 17 %.
1967	Reisby	Risk to children of schizophrenic parent 7 %.
1967	Zerbin-Rudin	Summarized family data of previous 50 years. Risk to children 12 %. Risk to full sibs 8.5%. Risk to parents 5 %. Aunts and Uncles 2 %. Nieces and nephews 2 %. First cousins 3 %.
1968	Erlenmeyer	Summarized all previous studies of offspring of two schizophrenic parents. Minimal estimated risk 32 %.
1980	Scharfetter	Starting with 269 probands diagnosed using ICD-9 criteria, 49 of 726 first degree relatives were identified as being affected with schizophrenia. Prevalence risk of 7 %. Morbid risk of 9 %.
1980	Tsuang	Starting with 200 probands diagnosed by clinical consensus, 20 of 375 first degree relatives were identified as being affected with schizophrenia. Morbid risk of 5 %.
1982	Pope	Did not find increase if strict diagnosis is used. This is the only modern study that failed to find increased risk.
1983	Guze	Starting with 44 probands diagnosed using Feigner criteria, 4 of 111 first degree relatives were identified as being affected with schizophrenia. Prevalence risk of 3.6 to 4.5 % depending on strict or broad diagnoses.
1985	Baron	Starting with 90 probands diagnosed with DSM III criteria, 19 of 366 first degree relatives were identified as being affected with schizophrenia. Morbid risk of 5.8%.
1985	Kendler	Starting with 332 probands diagnosed with DSM III criteria, 26 of 723 first degree relatives were identified as being affected with schizophrenia. Morbid risk of 3.7%.

Implications of Family Studies in Schizophrenia

There are some general conclusions that can be derived from family studies of schizophrenia. First, the risk of schizophrenia in the general population ranges from about 0.5 to 1.5 percent depending upon the study. However, the consensus is that about 1 percent of the population has some form of schizophrenia. The risk of schizophrenia to offspring if one parent is affected is about 10 percent or ten times higher than the general population. The cumulative risk of schizophrenia in first degree relatives of a schizophrenic proband is from 5 to 15 percent. In schizophrenia, we see most of what you would expect to see if a trait was genetic. For instance, it is highly familial and the risk seems to be graded by the distance of relationship to the proband, and blood related family members have a far higher risk than family members that are not biologically related (spouses and step siblings). If schizophrenia were totally environmental, like a viral epidemic, there would be far less gradation of risk and far less familiality, and a near equal risk to spouses and step siblings.

It is important to realize that family studies can only play a very limited role in psychiatric genetics. They do not prove that genetic factors play a part in a disorder, although they can support the idea. As has often been pointed out, many things are familial, but not genetic. For example, infections can be familial but are not primarily genetic. This argument has often been used to discount the role of family studies. In summary, family studies support the idea that schizophrenia is a genetically transmitted disorder, but do not prove it. As a result, family studies in schizophrenia can be performed until the end of time and it will not bring us one step closer to the real goal of finding a gene or genes for schizophrenia. Given the limited conclusions derived from schizophrenia family studies, one cannot help but share the sentiments of one of the pioneers on the subject – Eugene Bleuler:

> Forty years after having started my research, I was faced with a large number of . . . data on the families of 627 schizophrenic patients . . . , 1,117 of their fathers and mothers, on their 2,048 siblings and on over 5,000 of their other relatives . . . My juvenile hopes to find simple Mendelian ratios were disappointed like those of most psychiatrists . . . the results as regards heredity were modest. They permitted the statement . . . the development of schizophrenic psychoses – were due to the interaction between inborn dispositions and environmental influences – a vague statement that doesn't mean much . . .

Confronted with these modest results, genetic research became uninteresting to many psychiatrists – and psychiatry to many geneticists.

Twin Studies in Schizophrenia

Twin studies have a long and rich history in the study of schizophrenia. They have been analyzed and reanalyzed more that any other type of study. They are a powerful tool to answer the question of whether genetic factors are important in schizophrenia, but they do little else. They cannot, for instance, say much about the mode of transmission, and they certainly contribute nothing to the localization of specific genes. Nevertheless, for many years they were one of the few accessible ways of adding to the knowledge base about psychiatric genetics, and many of the most important pioneers in psychiatric genetics spent most of their productive lives in pursuit of them. For that reason alone, it is appropriate to briefly review the major schizophrenia twin studies and to derive whatever conclusions that can be deciphered from them.

Luxenburger Study (1928 - 1936) - The first systematic use of the twin study technique to examine the genetics of schizophrenia was performed by Hans Luxenburger in Munich, Germany. Using data from the Munich Psychiatric Clinic and Bavarian mental hospitals, he was able to amass a small number of MZ and DZ twins one or both of whom were affected with schizophrenia. These studies were modern in that effort was made to obtain an unbiased series instead of relying on the error prone method of compiling case reports of twins. No definitive number was ever given for the number of MZ and DZ pairs resulting in confusion in that the same data was presented in four different papers. The MZ concordance varied from 64 percent (12/19) in the 1928 report to 33 percent in the 1934 report, finally arriving at 51 percent (11/19) in the final 1936 report. DZ concordance was always 0 percent (0/13).

Rosanoff (1934) - This study of twin data collected from the USA and Canada reported on greater numbers of twins than the Luxenburger data. However, it is not a particularly organized report and problems with diagnostic clarity and ascertainment make this a difficult report to interpret.

Essen - Möller (1941) - This small but very interesting

study based on consecutive admissions to Swedish mental hospitals is noteworthy for its long follow-up. Only five of 11 MZ pairs were originally concordant, but on follow-up thirty years later two additional pairs had become concordant. The tendency for MZ concordance to increase over the life of the MZ pair had been speculated by a number of investigators, but this report contains the best evidence that it actually occurs.

Kallmann (1946) - This report by Franz Kallmann based on patients from the New York area was by far the most comprehensive up to that time. The strong support for the genetic basis of schizophrenia drew massive criticism from those inclined to favor non-genetic models. Some questions of ascertainment bias based on excess of MZ pairs appears unjustified. Others have questioned the diagnostic procedure implicating differences in how MZ and DZ pairs were diagnosed. Due to varying interpretations, the MZ concordance rate has ranged from 50 to 86 percent and the DZ concordance has varied from 6 to 15 percent. The higher concordance rates for MZ twins have been attributed to many factors. The only one that is scientifically valid is that he drew from a patient population that was biased for more severely ill individuals.

Slater (1953) - This report was based on retrospective ascertainment of twin probands from London area mental hospitals. This study was notable for the inclusion of specific patient data in the final publication, thus allowing others to analyze the data in novel ways.

Inouye (1963) - The importance of this study is primarily due to it being the first study based on a non-Caucasian population. Otherwise, it was flawed from ascertainment bias in that the twins were not from a continuous sample.

Tienari (1963-1975) - This study, based on population registries of twin births in Finland, was one of the most controversial of all the twin studies. In the initial 1963 report none of the 16 MZ twin pairs were concordant. This result seemed to provide evidence to support the critics of genetic studies. The study, based only on males, has been ongoing for a number of years. As the sample size increases and the subjects increase in age the concordance rates are increasing. In addition, the authors acknowledged using very strict diagnostic criteria. If less restrictive criteria were used it increased the MZ concordance considerably.

Kringlen (1964-1968) - This study was based on data collected from the Central Registry for Psychosis in Norway. The cases were reported in detail. The data varied slightly over a four year period, but generally showed slightly less MZ concordance than previous studies.

Harvald & Hauge (1965) - This study was based on all twins born in Denmark from 1870-1910. Although the authors offered some disclaimers because the diagnoses were often based on old medical records, they nevertheless found a MZ concordance of 44 % (4 of 9) and a DZ concordance of 12 % (4 of 33).

Fischer et al. (1969) - This study was based on the same national twin register in Denmark. It generally supported the conclusions from the previous study and found higher MZ concordance and DZ concordance rates in a larger sample.

Pollin et al. (1969) - This study was based on a National Academy of Science and National Research Council Twin Panel of United States veterans from 1941-1955. As such, it is a unique population not representative of the general population. Zygosity was not reliably determined, nor were the diagnostic criteria uniform. As a result the MZ concordance rate is quite low. One interesting item from this study is the ability to calculate schizophrenia prevalence rates. The prevalence of schizophrenia in MZ twins was 91/9426 = .0097 while the DZ twins was 152/12480 = .012. Thus proving that monozygosity does not increase the likelihood of developing schizophrenia as had been promoted by some wishing to minimize the role of genetics in schizophrenia. The data from the NAS-NRC study was reanalyzed and updated by Allen et al. in 1972 with an increase in the MZ concordance mostly due to a more careful analysis of military records. Finally, in 1983 Kendler et al. updated the twin data once more resulting in a further slight increase in MZ concordance.

Gottesman and Shields (1972) - This study supports the previous findings of schizophrenia twin studies. A MZ concordance of 58 percent and a DZ concordance of 12 percent were reported.

Mitsuda (1972) - This report was based on 27 twin pairs collected unsystematically. They reported a MZ concordance of 54 percent and a DZ concordance of 0 percent.

The results of these studies are listed in Table Five.

TABLE FIVE:
LIST OF SELECTED TWIN STUDIES IN SCHIZOPHRENIA *

#	YEAR	AUTHOR (COUNTRY)	Concordance # MZ N	MZ %	DZ N	DZ %
1)	1928-1936	Luxenburger (Germany)	11/19	**51**	0/13	**0**
2)	1934	Rosanoff (USA)	25/41	**61**	7/53	**13**
3)	1941-1970	Essen-Möller (Sweden)	7/11	**64**	4/27	**15**
4)	1946	Kallmann (USA)	120/174	**69**	34/296	**11**
5)	1953	Slater (England)	24/37	**65**	8/58	**14**
6)	1961	Inouye (Japan)	33/55	**60**	2/11	**18**
7)	1963-1975	Tienari (Finland)	7/21	**33**	6/42	**14**
8)	1964-1968	Kringlen (Norway)	31/69	**45**	14/96	**15**
9)	1965	Harvald and Hauge (Denmark)	4/9	**44**	4/33	**12**
10)	1969	Fisher et al. (Denmark)	14/23	**61**	12/43	**28**
11)	1969	Pollin et al. (USA)	11/80	**14**	6/146	**4**
12)	1972	Allen (Reanalysis of Pollin)	26/95	**27**	6/125	**5**
13)	1983	Kendler (Reanalysis of Pollin)	60/194	**31**	18/227	**6**
14)	1972	Gottesman et al. (England)	15/26	**58**	4/34	**12**
15)	1972	Mitsuda (Japan)	14/26	**54**	0/1	**0**

* For studies reported in multiple reports, the most comprehensive report is shown.
# All analyses are probandwise unless noted.

Monozygotic Schizophrenic Twins Reared Apart

The weight of evidence that MZ twins are much more prone to developing schizophrenia when a co-twin is affected (versus DZ co-twins) is overwhelming. However, some critics could not accept these findings and offered the explanation that being a MZ twin makes a person more susceptible to developing schizophrenia if the other twin has schizophrenia. This argument seems somewhat fanciful at best, but it has been answered by collecting data on identical twins with schizophrenia that were reared apart – no mean feat considering that the prevalence of schizophrenia is 1 percent, the prevalence of twining is about 1 percent, and that twins reared apart are rare. The results are presented in Table Six.

TABLE SIX:
SCHIZOPHRENIC MONOZYGOTIC TWINS REARED APART.*

	Year	Author	Concordant	Discordant
1)	1938	Kallmann	1	0
2)	1941	Essen-Moller	1	0
3)	1956	Kallmann and Roth	1	0
4)	1962	Shields	1	0
5)	1963	Tienari	0	2
6)	1967	Kringlen	1	1
7)	1972	Mitsuda & Inouye	8	4
		Total	13	7

* Modified from Liston and Jarvik (1976)

The results show a MZ concordance of 65 percent (13/20), which is similar to the average of all the MZ concordance rates.

Use of Twin Studies to Establish Diagnostic Criteria

In 1984, McGuffin et al. used a defined set of MZ and DZ twins as a tool to determine the "best" schizophrenia diagnostic criteria. This paradigm was based on the supposition that the criteria that made the largest difference in MZ-DZ concordance rates is the closest to the actual pathology of schizophrenia. This method clearly helped evolve the modern diagnostic criteria for schizophrenia.

Schizophrenia Rate in the Offspring of Discordant Monozygotic Twins

Another interesting study by Gottesman et al. (1987) suggests that although the monozygotic unaffected co-twin of a schizophrenic proband may be schizoid or normal, when the offspring of both groups are analyzed the rate of schizophrenia is similar in both. This adds further evidence that the genetic diathesis in schizophrenia may not be fully penetrant (even in MZ twins).

Conclusions From Schizophrenia Twin Studies

The twin data described here has been used over and over to support or discredit various theories. They have been analyzed and reanalyzed looking for reasons why they are flawed or over or under estimating the inheritance of schizophrenia. During the height of the anti-genetic period in psychiatry they were set upon with zeal and ridicule. However, they still speak for themselves. In over 15 studies, all but one of which confirms that genetic factors are undeniably important in schizophrenia. Even allowing for diagnostic, ascertainment and statistical problems, monozygotic concordance rates are consistently 3 to 5 times higher than dizygotic concordance rates. This translates to a 40 to 60 times higher risk for a co-twin than the general population. Twin studies are a tremendous tool in psychiatric genetics, but the schizophrenia studies shown here point out their biggest limitation. Other than proving that a strong genetic component exists, they do not provide much more solid information.

Adoption Studies in Schizophrenia

Adoption studies in schizophrenia evolved to answer one critical question. If schizophrenia is a familial condition, is it the biological (genetic) influence which determines transmission or is it the rearing environment. Although twin studies went a long way toward answering this question, only adoption studies could clearly answer it. The first adoption study of schizophrenia was performed by Heston in 1966 and a number of important studies followed (Table Seven). Every adoption study has either supported the biologic influence over the role of environment in schizophrenia or has at least not indicated any strong role for environment. Adoption studies in schizophrenia have answered the question. It is the biologic factors that are transmitting the schizophrenia phenotype between generations.

TABLE SEVEN :
SELECTED ADOPTION STUDIES IN SCHIZOPHRENIA

Date	Author	Findings
1966	Heston	Compared the rate of schizophrenia in the adopted offspring (separated at birth) from schizophrenic mothers (5/47 = 10.6 percent) to a control group (0/50 = 0 percent). It is important to note that the 10.6 percent rate in children was similar to the rates in family studies although the children were not raised by the biologic parent. Also the lack of any schizophrenic offspring in the control group speaks strongly against the argument that the stress of adoption "creates" schizophrenia.
1966	Karlsson	Compared the rate of schizophrenia in the biological sibs of adopted schizophrenic probands and found a 21 percent rate (6 of 29) versus the foster sibs of the adopted schizophrenic probands and found a 0 percent rate (0 of 28). It is unclear why the biologic sib rate is so high (twice the family study rate, but may be due to the fact that it was done in Iceland which may have unique genetic founder effects).
1971	Rosenthal	Compared the rate of "schizophrenia spectrum disorders" in the offspring (separated at birth) from schizophrenic mothers (24/76 = 31.6 percent) to a control group (12/67 = 12.9 percent). High risk rates are due to use of less stringent diagnostic criteria and inclusion of spectrum disorders.
1971	Kety, et al.	Compares the rate of schizophrenia among biological relatives with that of adoptive relatives. The results show much higher rates in biological relatives.
1976	Higgins	Compared the rate of schizophrenia in 25 adopted away children of schizophrenic mothers and 25 control children born and raised by schizophrenic mothers. Rates of 32 percent (8/25) in adopted verses 16 percent (4/25) in non-adopted are not statistically significant due to small number studied.

However, like twin studies, that is the end of the story. Adoption studies cannot precisely define the genetic role or bring us closer to a specific gene or genes.

Attempts to Define a Mode of Transmission in Schizophrenia

Segregation analysis has not resolved the mode of transmission for schizophrenia. Theories range from dominant and recessive single gene models to polygenic models (Table Eight), but the evidence does not conclusively support any mode of transmission (Kidd & Cavilli-Sforza, 1973). Genetic heterogeneity best explains the inconsistent data obtained from segregation analysis. Heterogeneity implies that the schizophrenia phenotype may be produced by more than one genetic defect, either from different enzyme mutations in the same biochemical sequence, different genes in early brain development, or from separate defects in unrelated pathways. It is not surprising that there appears to be genetic heterogeneity in schizophrenia since heterogeneity is common in many genetic disorders that have a prevalence 5 to 100 times less than schizophrenia.

TABLE EIGHT:
THEORIES OF GENETIC TRANSMISSION IN SCHIZOPHRENIA

Year	Investigator	Theory of Transmission
1916	Rüdin	Two recessive genes
1938	Kallmann	One gene recessive
1946	Kallmann	One gene recessive
1953	Book	One gene recessive
1953	Book	One gene recessive
1958	Slater	One gene
1960	Edwards	Polygenic
1962	Garrone	One gene
1964	Birch	Two gene
1964	Karlsson	Two genes
1966	Erlenmeyer	Heterogeneity

TABLE EIGHT: (CONT)

1967	Gottesman	Polygenic
1968	Kety	Polygenic
1968	Kringlen	Polygenic
1970	Heston	One gene
1970	Elston	One gene
1973	Kidd	Data supports boths single gene and polygenic models
1980	Crow	Speculates on two types
1982	Baron	One gene
1982	Orourke	One gene
1983	Farmer	Cluster analytic approach
1985	Mcgue	One gene
1986	Eaves	Describes model
1984	Dworkin	Heterogeneity
1985	Crow	Review of two system concept
1987	McGuffin	Heterogeneity. Poor genotype/phenotype relationship

As can be seen in this table, little is known about the genetic mode of transmission of schizophrenia, and little will remain known until breakthroughs are made in the molecular genetic study of schizophrenia. In the mean time, little is to be gained by endlessly speculating on the merits of the various models.

Conclusion

Schizophrenia is clearly caused, at least in part, by genetic factors. The classic approaches of family, twin, and adoption studies have contributed greatly to our knowledge of schizophrenia, but they have not and cannot answer the more important questions that are now being asked. Only a molecular genetic approach holds the promise of localizing genes involved in its transmission.

THE GENETICS OF BIPOLAR DISORDER (MANIC-DEPRESSIVE ILLNESS)

Affective disorders are characterized by a disturbance in mood. The disruption can come in the form of elevated mood (mania) or depressed mood (depression). Bipolar disorder is a disturbance of mood characterized by periods of elevated mood and periods of depressed mood, although numerous variations of this disorder exist. Numerous endocrine and neurologic ailments can effect mood. The focus in psychiatry is on disorders of mood not attributable to known medical illness. The genetic study of affective illness required a clear delineation from other forms of mental illness. The work of Emil Kraepelin at the turn of the century was mostly responsible for separating affective disorders from other disorders, such as schizophrenia.

Family Studies

Not surprisingly, the first studies of the genetics of affective disorders occurred in the 1920's by students of Kraepelin. However, from the 1920's until the mid 1960's, manic-depression (bipolar disorder) was not clearly separated from pure depression (unipolar disorder). The separation of the two disorders was the result of the work of Leonhard and Perris in Berlin in the early 1960's. Studies since then have placed a prevalence rate of about 1 percent for bipolar disorder (Zerbin-Rudin). In the 1980's bipolar disorder was separated into bipolar type I (classic manic-depression) and bipolar type II (depression with periods of hypopmania only). Genetic studies of affective disorders conducted before 1960 suffer from this conceptual flaw, however they are listed in Table Nine along with the more modern studies.

TABLE NINE:
SELECTED FAMILY STUDIES IN BIPOLAR DISORDER

Year	Investigator	Findings
1921	Hoffman	First family study of bipolar disorder. Overly inclusive, used small numbers.

TABLE NINE: (CONT)

1923	Rüdin	Second study, although performed prior to Hoffman.
1929	Banse	Studied cousins of individuals with manic-depressive illness. Only 40 percent of participants were interviewed directly. Risk of 3.5 percent to cousins.
1932	Luxenburger	Previous four studies reviewed by Luxenburger.
1936	Slater	Fifth study. Established most exhaustive criteria for study. Reviewed all previous studies.
1936	Roll	Increased familial risk.
1936	Weinberg	Increased familial risk.
1938	Slater	Reviewed and refined data.
1952	Stenstedt	Ratio's of risk consistent with dominant inheritance
1965	Winokur	Increased familial risk.
1966	Angst	Studied unipolar and bipolar separately.
1967	Winokur	Increased familial risk.
1982	Weissman	Increased familial risk.
1983	Egeland	Identified large family in Amish population.
1983	Winokur	In this summary paper, found 847 first degree affected relatives of bipolar patients. Looked at first degree relatives of bipolar I and bipolar II patients and found bipolar disorder in 8.3 percent (4.9 percent bipolar type I, 4.4 percent bipolar type II), cyclothymia in 3.0 percent, and schizoaffective disorder in 1.2 percent.

Twin Studies in Bipolar Disorder

Twin studies in bipolar disorder were mostly done before unipolar and bipolar were separated, which makes interpretation very difficult. The first seven studies listed in Table Ten are virtually useless for this reason. They are listed for purposes of inclusion only. The most important studies are the final four, which were all done with modern diagnostic criteria. The pooled data from these last four studies suggests about 65 percent MZ concordance and 14 percent DZ concordance. This is higher than schizophrenia and represents a very high heritability for bipolar disorder. The question of shared environment was addressed by Price, who reviewed the available literature and found 12 cases of MZ twins reared apart where one co-twin has bipolar disorder. Eight of these twelve were concordant for bipolar disorder, which roughly matches the concordance rates above. Again, however, the finding only supports the contention that the bipolar phenotype has a strong genetic influence and can provide little else.

TABLE TEN:
TWIN STUDIES OF BIPOLAR DISORDER*

#	YEAR	AUTHOR	Concordance #			
			MZ		DZ	
			N	%	N	%
1)	1930	Luxenburger (Germany)	- /-	-	-/-	-
2)	1932	Luxenburger (Germany)	- /-	-	- /-	-
3)	1935	Rosanoff et al. (USA)	- /-	70	- /-	16
4)	1941	Essen-Moller	- /-	-	- /-	-
5)	1953	Slater (England)	- /-	-	- /-	-
6)	1954	Kallman (USA)	25/27	92	13/55	24
7)	1959	De Fonseca	- /-	71	- /-	39
8)	1965	Harvald et al. (Denmark)	- /-	50	- /-	3
9)	1967	Kringlen	- /-	33	- /-	0
10)	1977	Bertelsen	- /-	58	- /-	17
11)	1986	Torgersen (Norway)	- /-	75	- /-	0

*Prior to 1965 no distinction made between bipolar and unipolar.

Mode of Transmission in Bipolar Disorder

The exact mode of transmission of bipolar disorder is not known. In fact, a definitive mode of transmission to explain all cases of bipolar disorder is extremely unlikely. Bipolar disorder is almost certainly genetically heterogeneous. This uncertainty about mode of transmission has not stopped investigators from speculating, however, as evidenced by 70 years of theories. Many large bipolar pedigrees are consistent with autosomal dominant transmission. These families show multiple affected members segregating through several generations. Further evidence for this theory includes the fact that family studies show that the risk of parents for bipolar disorder is similar to the risk of the children. Slater postulated an autosomal dominant mode of transmission as early as 1936. Autosomal recessive inheritance appears unlikely as an explanation for most pedigrees of bipolar disorder. The presence of ubiquitous multiple generation families is one indication. The lack of increased risk with consanguinity is another. Sex-linked dominant transmission has been postulated by Winokur in 1969 due to the reported 2 to 1 ratio of affected males to females and the presence of reports of X linkage in some families. However, ample evidence for male to male transmission makes this a partial explanation at best. Sex linked recessive transmission is unlikely due to lack of studies showing excess males over females. Currently most models emphasize polygenic inheritance.

TABLE ELEVEN:
THEORIES ON THE MODE OF TRANSMISSION OF BIPOLAR DISORDER.

YEAR	AUTHOR	TYPE	THEORY
1921	Hoffman	Polygenic	Three nonspecified genetic factors.
1923	Rüdin	Polygenic	One autosomal dominant, two autosomal recessive.
1932	Luxenberger	Polygenic	One autosomal dominant , one autosomal recessive.

TABLE ELEVEN: (CONT)

1935	Rosanoff	Polygenic	One autosomal dominant, and one X-linked dominant factor.
1936	Slater	Monogenic	Autosomal dominant with reduced penetrance.
1937	Lenz	Monogenic	Autosomal dominant.
1954	Kallmann	Monogenic	Autosomal dominant with reduced penetrance.
1969	Winokur	Monogenic	X-linked dominant
1971	Slater	Polygenic	Multiple genes.
1971	Perris	Polygenic	Multiple genes.
1972	Mendlewicz	Monogenic	X- linkage

Heterogeneity is likely to be very important in bipolar disorder. Different genetic subtypes of bipolar may have different modes of transmission. Further progress in this area is dependent upon advances in the molecular studies of bipolar disorder and cannot be resolved through further family, twin or adoption studies.

UNIPOLAR DISORDER (MAJOR DEPRESSION)

> "I have often heard him lament that he inherited from his father a morbid disposition both of body and mind - a terrifying melancholy, [of] which he was sometimes apprehensive, bordered on insanity."
>
> A contemporary describing
> Samuel Johnson (1709-1784)

Unipolar disorder was not clearly differentiated from bipolar disorder until 1966. For this reason, there are only a few studies of pure unipolar disorder. However, since all these studies have been done in the modern era, they have few major systematic problems. Several family studies of unipolar disorder have been done. In 1980, Tsuang et al. reported a study where 362 at risk relatives were identified from unipolar probands. Of these 362, sixty-six were ill with unipolar disorder. This risk of 18 percent was more than double the control group of surgical patients who had an eight percent risk. In 1981, Mendlewicz et al. presented a unipolar disorder family study. Beginning with a sample of 106 probands diagnosed with unipolar disorder, the sample was further divided into early and late onset with the division at age 40. They found that 53 of 242 (21.9%) first degree relatives of the early onset group were affected while only 32 of 228 (14.0%) of the late onset group were affected. This suggests that there are two subtypes of unipolar illness, an early form that is more genetic and a late onset form that is less genetic. Also, the early group had an excess of females compared to the later group. However, both groups had a much higher risk than general population estimates.

One major twin study has been completed that separated out unipolar from bipolar disorder. The Danish twin study by Bertleson et al. in 1977 reported on 44 pairs of twins taken from a systematic ascertainment. MZ concordance was 43 percent while DZ concordance was 19 percent. Adoption studies in unipolar disorder have not been done. One of the major problems in studying the genetics of unipolar depression is that the diagnostic criteria are not focused enough to clearly differentiate unipolar depression from mild forms of situational depression. This tends to dilute the role of genetics in this disorder. Most investigators have recognized this and tend to study other less complicated phenotypes.

PRIMARY DEGENERATIVE DEMENTIA OF THE ALZHEIMER TYPE

Primary degenerative dementia of the Alzheimer type is the most common cause of dementia. More than 1.5 million people in the United States have Alzheimer disease and 100,000 die of it each year. Alzheimer disease was first described in 1906 by Lois Alzheimer. Alzheimer disease is often mistakenly considered to be synonymous with dementia. In fact, there are over 50 other documented causes of dementia including metabolic, carcinogenic, and structural abnormalities (hydrocephalus). Dementia is even reversible in up to 10% of cases. However, Alzheimer disease is the most common form and is not reversible although there is some variability in the rate of progression. Alzheimer disease has distinct histopathologic features. Brains of patients with Alzheimer disease contain neuritic or senile plaques, neurofibrillary tangles, and granovacuolar neuronal degeneration. The diagnosis is made clinically, but only autopsy or brain biopsy is definitive. Structurally, there is often shrinking of brain mass and enlargement of ventricles on CT scan. EEG often shows diffuse slowing.

Anecdotal reports of familial clustering of dementia proceeded any scientific attempts to analyze the disorder. In addition, there are no shortages of very large families that have been reported to have very high incidences of dementia. One family revealed dementia in eleven generations (Foncin, 1986). A number of family studies have been performed as shown in Table Twelve.

TABLE TWELVE:
SELECTED FAMILY STUDIES IN ALZHEIMER DISEASE

Year	Investigator	Findings
1952	Sjogren	Found a 4.6 percent risk in first degree relatives of 36 probands. No control group was used.
1962	Constantinades	Found a 2.3 percent risk in first degree relatives of 97 probands. No control group was used.
1981	Heston	Found a 9.9 percent risk in first degree relatives of 125 probands.
1983	Heyman	Found a 5.1 percent risk in first degree relatives of 68 probands.
1984	Wright	Found 80 percent heritability for early onset cases.
1987	Mohs	Found the highest risk of any study so far, 13.9 percent risk to first degree relatives.

Only one systematic twin study has been done (Nee et al., 1987), and the very small numbers (17 MZ and 5 DZ pairs) severely limits the interpretation. However, MZ concordance was 41 percent and DZ concordance was 40 percent. No adoption studies have been done, probably because nobody suggests that dementia is caused by how a person is raised. Most of the individual families reported have appeared to have autosomal dominant transmission. There has been speculation that there are differences between early onset and late onset forms of dementia, with the early onset forms being more familial than the late onset forms. Also, it has been speculated that there are genetic differences between sporadic and familial cases (Pratt, 1970). Like the other broad behavioral phenotypes, heterogeneity is very likely (Bondareff et al., 1987). Segregation analysis of 410 consecutively ascertained nuclear families by Rao, et al. (1994) revealed a complex inheritance that rejected a single gene model or random environmental transmission, but favored a model of genetic heterogeneity.

ALCOHOLISM

Alcohol dependence is defined as the excessive use of alcohol in a pattern that ultimately leads to relationship problems, legal problems, health problems, and/or work problems. It is not possible to define alcoholism on the basis of amount, frequency, or duration of alcohol use. Some studies suggest that around five percent of males and 1 to 2 percent of females are alcoholics during their lifetimes. Something as volitional as alcohol use would on the surface seem unlikely to be strongly influenced by genetic factors, but it has long been noted that a family history of alcoholism is the strongest predictor of an individual's risk for alcoholism. Family studies of alcoholism are consistent in showing an increased relative risk for all family members. Some studies suggest first degree male relatives of alcoholics have as high as a 50 percent lifetime risk of alcoholism.

Twin Studies in Alcoholism

Kaij (196) presented a twin study of alcoholism using modern criteria. Swedish twin registries were used to pick out 174 twin pairs for study. Kaij ranked each subject using a five scale of severity. MZ twin concordance was very high – 85 percent. However, so was DZ twin concordance – 67 percent. Kaij also found evidence to suggest that concordance increased as severity of alcohol use increased suggesting genetic factors may play a larger role in the more severe forms of alcohol abuse.

Adoption Studies of Alcoholism

The Danish adoption study paradigm was also applied to the question of alcoholism. Several strategies were employed. Goodwin et al. (1973) studied 55 men (age 23-45) who had a biological father who was alcoholic, but who were placed for adoption before six weeks of age. They used a control of similar men with no history of alcoholism in their fathers. The results showed that alcoholism occurred in about 20 percent of the study group and only 5 percent of the control group. Goodwin et al. (1974) looked at adopted away brothers versus non-adopted brothers and found little difference in the two groups in the rate of alcoholism, which implies that for all the talk about the effect of adoption it seemed to have had no effect in this case. A third study by Goodwin et al. (1977) involved the use of 49 adopted away daughters of alcoholic fathers with a control of 47 adopted away daughters of non-alcoholic fathers. The results showed no significant difference in alcoholism between the groups, which is most likely due to small sample size given that the female alcoholism rate is around one percent.

STUTTERING

The definition of stuttering is deceivingly straightforward: "a disorder in the rhythm of speech in which the individual knows precisely what he wishes to say, but at the same time is unable to say it because of an involuntary repetition, prolongation, or cessation of sound (World Health Organization, 1977)." Moderate to severe stuttering is as easy to recognize as it is to define. Mild stuttering, in contrast, can be quite difficult to recognize and leads to a decreased inter-rater reliability. The prevalence of stuttering varies with age. In preschool age children, so-called developmental dysfluency may be as high as 3-5 percent, whereas in the adult population the general prevalence is closer to 0.7 to 1.0 percent. This relatively high prevalence in children from the time they first speak to when they start school is felt by many speech-language pathologists to reflect a mixture of transient developmental dysfluency (which may be self limited and not reflect any underlying neurophysiological deficit) and a less prevalent persistent developmental dysfluency, which does reflect an underlying speech motor dyscontrol (Seider 1983). Stuttering starts very early, usually between 3-5 years of age. Over half of the persistent developmental stutterers are recognized to be stutterers at the earliest onset of speech. Virtually all persistent stutterers are recognized as such by 8 to 12 years of age. Like many behavioral disorders of childhood there seems to be a distorted male-to-female affection ratio, usually in the range of 3 to 4 affected males to one affected female.

Family Studies of Stuttering

The familial nature of stuttering was first noticed in the 19th century and was first studied in the early 20th century (Van Riper, 1982). Since 1914, family studies have repeatedly shown that the risk of stuttering among first degree relatives is increased several times that of the general population, as is shown in Table Thirteen.

TABLE THIRTEEN:
FAMILY STUDIES OF STUTTERING

Date	Investigator	Findings
1914	Makuen	Familial tendency.
1917	Bryant	Familial tendency.
1924	Gutzman et al	Found that 27 percent of stutterers have a positive family history.
1937	Bryngelson	Positive family history 2.5 times greater in stutterers versus controls.
1939	Wepman et al	Positive family history 4 to 6 times greater in stutterers versus controls.
1939	West et al	Positive family history 3 times greater in stutterers versus controls.
1939	Nelson	Ascertained 204 stutterers (163 males; 41 females) and 204 age and sex matched controls. Then compiled family histories by a modified family history method using some other family members

TABLE THIRTEEN: (CONT)

		along with the probands to complete the pedigrees. Found that 100 of the 204 stutterers had a strong positive family history, while only 4 of the 204 controls had a strong positive family history. This is a 25 fold risk.
1942	Johnson et al	Positive family history 4 times greater in stutterers versus controls.
1945	Meyer	Positive family history 10 times greater in stutterers versus controls. Study of 116 stutterers (103 males and 13 females) ascertained from the National Hospital for Speech Disorders in New York City. Found that 61 percent of stutterers had a family history of stuttering while only 6.5 percent of a control group of non-stutterers did. He also found that 25 of the 61 stutterers had at least one parent that stuttered.
1977	Sheehan, et al.	Reviewed all family studies and found " those who have specifically studied familial occurrence in stuttering have uniformly found far greater incidence of stuttering relatives in the families of stutterers."
1984	Kidd et al.	Reviewed data from Yale Family Study of Stuttering The results from a study of over 600 families showed that the incidence of stuttering is increased up to 30 times that of the general population in first degree relatives of adult stutterers.

The total risk to first degree relatives is about 25-35 times the general population risk. Like many behavioral disorders, phenocopies are likely to be present in stuttering. An interesting approach to this problem was conducted by Sheehan and Martyn. They analyzed the spontaneous recovery rate in stutterers with a family history versus those without (the assumption being that stutterers without a family history are more likely to be phenocopies). They found that the stutterers with a family history showed less spontaneous recovery and concluded that "the results support the interpretation that there may be either a familial or genetically predisposed subtype of stutterer whose disorder is less amenable to environmental changes."

Twin Studies in Stuttering

A number of anecdotal twin studies have been reported in stuttering. In 1981, Howie et al. collected the twin studies previously published and presented new evidence from a systematically obtained sample of 30 twins (Table Fourteen). They found that the risk of stuttering in a co-twin of a MZ twin is 63 percent and in a DZ twin is 23 percent.

TABLE FOURTEEN:
TWIN STUDIES IN STUTTERING

Reference	MZ TWINS Pairs #	MZ TWINS Concordant #	MZ TWINS Concordant Rate	DZ TWINS Pairs #	DZ TWINS Concordant #	DZ TWINS Concordant Rate
(1945-81) Case Reports*	50	39	0.78	78	7	0.09
(1981) Howie et al	16	10	0.63	13	3	0.23
Total #	66	49	0.74	91	10	0.11

* = compiled by Howie et al. from 10 individual reports.
\# = this calculation in based on raw data and is not corrected.

A more recent twin study using a twin registry in Australia found similar results in an unselected population revealing a MZ/DZ pairwise concordance ratio of 5.7:1 Thus, the results of twin studies in stuttering suggest that the familiality of stuttering is due in large part to genetic factors.

Mode of Transmission of Stuttering

Early in the study of the genetics of stuttering several investigators asserted that the stuttering phenotype was due to single gene transmission. Wepman (1939) suggested stuttering was a recessive trait, or because of the increased male-to-female ratio, a sex linked trait. Meyer (1945) argued against the assertion of recessive inheritance by pointing out that in his series of 52 patients with a positive family history, he found evidence of stuttering on both sides of the family in only 5 families. He argued that a recessive trait would clearly have more evidence of bilineal descent. He further argued that since 25 of 61 stutterers had at least one parent with stuttering, that this further argued against recessive inheritance. Meyer then continued the argument against the concept that stuttering was due to either a recessive or dominant sex-linked trait. For a recessive sex-linked trait he pointed out that the hemizygous males would predominate the individuals with stuttering, while homozygous recessive females would be very rare (perhaps on the order of 0.03%). This would predict that there would be about 50 times as many male stutterers to female stutterers. This contrasted with the 4:1 male to female affected ratio observed. He admitted that heterozygous females that expressed the stuttering phenotype could narrow the ratio, but he argued that there should then be a less severe form of stuttering in affected females, which he did not find. He concluded from this that there was no evidence to support stuttering as a recessive sex-linked trait. He then explored the proposition that stuttering was a dominant sex-linked trait. He dismissed the possibility by noting that the daughters of affected males were far from 100% stutterers as would be expected from the transmission of a dominant sex-linked trait. Meyer then goes on to reject the idea of a dominant trait by implying that 50% of offspring do not have the trait.

There is currently no evidence for Mendelian single gene transmission in stuttering. For instance, Kidd et al. (1981) found by complex segregation analysis of stuttering using 386 probands with data derived on first degree relatives by the family history method that he could reject a single Mendelian locus as an explanation for all stuttering. Similar results were found by Cox (1984). Most investigators generally believe that stuttering is either a polygenic trait or a multifactorial trait.

One caveat to the conclusion that single genes do not play a role in stuttering is the potential role of heterogeneity. If many (perhaps dozens) of different single gene defects could lead to the phenotype of stuttering in a "final common pathway," then it is still quite possible, if not likely, that single gene effects can be found in stuttering. When this heterogeneity is coupled with factors that reduce the penetrance of the phenotype through chaotic influences of development, the resulting pedigree structures would be very similar to the pedigree structures that we now see for stuttering. Even the distorted male-to-female ratio can be explained under this scenario as simple sex-modified transmission (first suggested by West in 1939 and later expanded by Kidd in 1984).

TOURETTE SYNDROME

Gilles de la Tourette described the syndrome of chronic motor and vocal tics that bears his name in 1885. In addition, he was the first to note that the disorder has a familial tendency. The current diagnosis of TS requires the presence of chronic motor and vocal tics with an onset before age 21. The anatomic location, frequency, and severity of the motor tics can be highly variable. Vocal tics can take the form of barking, cursing or grunting, which can be very disabling to patients. The diagnosis of chronic tics (CT) is essentially the same as TS with the exception that *either* motor or vocal tics occur, but not both. Most investigators now believe that TS and CT are etiologically related with CT being a milder form of the disorder. The prevalence of strictly diagnosed TS is approximately 0.5 individuals per thousand population.

Since the first description of TS, anecdotal reports have implicated an increased rate of tic disorders in family members of patients. A number of studies over the past twenty years have provided further documentation of the familial nature of TS. The studies are quite variable in their methods of ascertainment and diagnostic criteria, so it is of limited use to try to determine specific rates of risk. One of the obstacles that TS researchers have encountered is the exact definition of the TS phenotype. The presence of a number of related disorders in family members of TS patients has spurred researchers to question if other phenotypes may be related to the underlying genetic defect. The best documented example is chronic tics (CT), which is felt to be a variant phenotype by most investigators. Another likely phenotype is obsessive compulsive disorder (OCD). Obsessional thinking was noted to occur in 30% of TS patients in a review by Fernando in 1967, and since then, several other investigators have provided additional evidence for the inclusion of OCD into the TS phenotype. The inclusion of OCD into the TS phenotype is not meant to imply that all cases of OCD are related to TS, but that one form of the behavior spectrum called OCD is related to TS.

The search for alternate phenotypes has created a controversy in the field of TS research. How the issue is resolved has implications for all of psychiatric genetics. Based on comparisons of the frequencies of various disorders in control groups versus relatives of TS patients, Comings et al. (1989) has shown evidence for an increased incidence of several disorders including attention deficit disorder, conduct disorder, phobic disorder, obesity, sleep disorders, and some forms of substance abuse. They suggest that these disorders are all alternative phenotypes of TS and that TS is one of "the most common genetic disorders of man." Other investigators take a more restricted view stating that although increased frequencies of other disorders do exist in TS relatives, the association is not genetic. They argue that the type of reasoning used is insufficient to claim cause and effect. If a disorder is increased in family members it could be caused by several other factors besides shared genetic etiology. One particular likely cause is assortative mating.

A single twin study has been reported in TS, and the results are suggestive of a major genetic factor. Price et al. studied 43 pairs of twins and found a MZ concordance rate of 53 percent and a DZ concordance rate of 8 percent using strict diagnostic criteria. Interestingly, when chronic tics were included as a phenotype, the MZ concordance rate rose to 77 percent and the DZ rate rose to 23 percent. With the increasing evidence for a strong genetic factor in the etiology of TS, a number of segregation analyses have been done. Four studies done between 1981 and 1984 support a single major locus model for TS, although polygenic models could not be ruled out. In 1986, a segregation study of TS was undertaken by Pauls and Leckman. The validity of this study was increased by the use of a direct interview approach where each family member was interviewed. The best fit for the results is with a single locus model, specifically autosomal dominant transmission. In addition, they are able to reject the "no transmission model" and the "polygenic model." An important point from the study is that, although strictly diagnosed TS fits the autosomal dominant model, the best fit was with TS, CT, and OCD as alternate phenotypes. The calculated penetrance using this model is 99 percent for men and 71 percent for women. There is a greater than 3 to 1 male to female ratio in strictly diagnosed TS. The difference in frequency among the sexes is probable not due to difference in transmission , but is more likely due to differences in sex specific expression. The ratio narrows considerable when OCD is included as a phenotype, suggesting that the gene expression is phenotypically more variable in women.

Hasstedt, et al. (1995) reported segregation analysis of a unique 182 member pedigree where TS is quite common. Simple Mendelian transmission was eliminated, but they found that a single locus could be responsible for the disorder if penetrance was 28 percent for heterozygotes and assortative mating in the 70-80 percent range was allowed. The evidence of genetic etiology in TS is convincing. The inclusion of CT and OCD as alternate phenotypes makes it even more convincing. When alternate phenotypes are included, evidence for a modified autosomal dominant transmission is strong. The possibility is likely, however, even with the apparent rarity of TS that heterogeneity will exist.

PERVASIVE DEVELOPMENTAL DISORDER (AUTISM)

Autism was first described as a syndrome by Kanner in 1943. It is a rare disorder with an approximate prevalence of 2 to 5 affected individuals per 10,000 children, and has a ratio of affected males to females of 3 to 1. In addition, there is an association with mental retardation with as many as 75 percent of autism patients having an IQ less than 70. Until recently, a major obstacle to understanding autism was the belief that autism is a childhood form of schizophrenia, a belief that is no longer endorsed. The diagnosis of autism involves observed deficits in four areas of function: (1) gross impairment in social interaction, (2) impaired verbal and nonverbal communicative skills, (3) impairment of imaginative skills, and (4) a marked restriction of activities and interest, which leads to stereotyped behavior and an intolerance of change in routine.

The question of familiality in autism has been answered by a series of studies, which reveal that autism is more common in siblings of probands than in the general population (Table Fifteen).

TABLE FIFTEEN:
AUTISM AMONG SIBLINGS OF AUTISTIC PROBANDS

Reference	# Families	# Siblings	# Autistic siblings
(1977) Folstein & Rutter*	21	36	1
(1980) Deykin & MacMahon	154	364	9
(1981) August et al	41	71	2
(1982) Minton et al	-	284	7
(1985) Ritvo et al*	40	80	2
(1985) Baird & August	29	51	3
POOLED DATA	285	886	24 (0.027)

(Table modified from Smalley et al.)
* Ascertained through twin studies.

The prevalence among siblings pooled from all the studies is approximately 270 affected individuals per 10,000, which is a 100 fold increase risk over the general population. Additional family risk studies involving parents or children are not possible in autism because of the virtual nonexistence of documented examples of procreation in autistic individuals. The implication of the family studies is that simple Mendelian models of dominant or recessive inheritance are unlikely as an explanation for the vast majority of cases of autism. This is apparent because only 3 percent of the siblings have the disease, far below the required 50 percent or 25 percent required for a dominant or recessive model respectively. In addition, the 3:1 male to female ratio is inconsistent with any autosomal model. X-linked recessive models are unlikely because the number of affected females is far in excess of expected.

Two systematic studies and a number of case reports comprise the twin study data in autism. Smalley et al. compiled single case reports of twins published between 1954 and 1985, and analyzed the other reports for bias (Table Sixteen).

TABLE SIXTEEN:
TWIN STUDIES IN AUTISM

Reference Concordant	MZ TWINS Total Pairs	Concordant Pairs	Rate	DZ TWINS Total Pairs	Concordant Pairs	Rate
(1954-85) Case Reports*	11	9	0.82	9	2	0.22
(1977) Folstein & Rutter	11	4	0.36	10	0	0.00
(1985) Ritvo et al.	23	22	0.96	17	4	0.24
Total	45	35	0.78	36	6	0.17

(Table modified from Smalley et al.)
* compiled by Smalley et al. from 15 individual reports.

They concluded that the "adjusted concordance" was 64 percent for MZ twins and 9 percent for DZ twins. The heritability index can be easily calculated [(64-9)/(100-9)] showing a value of 60 percent. This strongly implicates a genetic factor. Again, simple Mendelian models are not supported by the 9 percent DZ concordance rate, which is far less than the 50 percent or 25 percent expected in dominant or recessive models. The evidence for a significant genetic factor in the etiology of autism based on its current phenotypic description is, therefore, equivocal. This has led some investigators to question whether genetic factors play any role at all, and others to invoke complex multifactorial-polygenic models. To clarify the situation, some authors speculate that the autism phenotype is too narrow. They contend that when broadened to include minor cognitive and linguistic impairments the evidence for a genetic factor increases. August et al. found that if these mild cognitive dysfunctions were included, the prevalence of the disorder increased to 15.5 percent in siblings, while Baird et al. found 20 percent of siblings were affected with some degree of mild CNS dysfunction. In contrast, other authors have speculated that the phenotype of autism is over inclusive, in the sense that it does not differentiate between distinct forms caused by several possible genetic and environmental causes. Deykin et al., for instance, found many cases of autism were associated with a number of "unfavorable events in pregnancy," suggesting that many cases were caused by environmental factors. The etiologic heterogeneity may also reflect genetic heterogeneity. Numerous case reports have associated autism with various genetic disorders (Table Seventeen).

TABLE SEVENTEEN:
GENETIC SYNDROMES ASSOCIATED WITH AUTISM

Syndrome (modified from Reiss et al.)	Number of cases
XYY	3
XXYY	1
XXY	1
XXX	1
Fragile X	29
"Large" Y	12
Trisomy 21	3
Trisomy 22	1
Translocation 22/13	1
Gp+ (increased chromosomal material from a G chromosome)	1
3p- (deletion)	1
8p- (deletion)	1
13- (deletion)	1
17p- (deletion)	1
Phenylketonuria	66
Neurolipidosis	2
Hurler's Syndrome	2
Histidenimia	2
Oculocutaneous Albinism	1
Neurofibromatosis	2
Tuberous Sclerosis	7
Noonan Syndrome	1
Cornelia de Lange Syndrome	11
Williams Syndrome	2
Moebius Syndrome	1
Marshall Smith Syndrome	1

Reiss et al. reviewed these associations and concluded that "autism is a behaviorally defined phenotype which arises from diverse causes of central nervous system damage." This suggests that there may be several different types of genetically determined autism. Efforts to subdivide autism and to reduce the phenotypic overlap involve ongoing studies designed to find specific phenotypic determinants such as disease specific morphologic features, unique cognitive impairments, or specific biochemical markers.

Based on the theory of genetic heterogeneity, several studies have investigated the possibility that there are subsets of autism with strong genetic components. Baird et al. found evidence suggesting that severity of mental retardation may identify a subset of autism with a stronger genetic component. Ritvo et al. selected a subgroup of 46 families based on multiple cases of autism (two or more cases in one family), and determined that even with the strict phenotype, this subgroup had a sibling risk of 19 percent, which is much closer to the expected 25 percent of a recessive trait. However, the sex ratio was still distorted in this study, which is not easily explained. Another theory is that the families with affected females are a genetic subset, because the increased prevalence in males is due to the presence of phenocopies. As an example, fragile X may produce an "autistic like" phenotype, and 8 percent of male autistics have fragile X. Alternatively, it could be said that defects in the fragile X gene account for 8 percent of all autism leaving the other 92 percent to be accounted for. Several chromosomal abnormalities have also been associated with autism. It is clear that there is a genetic component to autism, however, significant questions regarding the exact role of the genetic factor or factors remain.

ATTENTION DEFICIT HYPERACTIVITY DISORDER (ADHD)

Various studies have shown an increased familial risk in ADHD. However, diagnostic issues are very complex in ADHD making family studies problematic. The diagnostic overlap with conduct disorder and oppositional defiant disorder contribute to the diagnostic uncertainty. One study of brothers of ADHD probands showed a 3 fold increased risk of ADHD. Another finding is that individuals with ADHD often grow up to have drug and alcohol problems as well as antisocial behavior. Full sibs of ADHD probands have a 5-6 times higher rate of ADHD than half sibs. An adoption study in ADHD was consistent with biologic transmission of ADHD, alcoholism, and antisocial personality disorder.

SPECIFIC DEVELOPMENTAL DISORDERS (LEARNING DISABILITIES)

Within the diagnosis of learning disability there are several specific disorders. They include academic, language, speech, and motor skill disorders. Major problems with diagnostic criteria exist for these disorders making genetic studies very difficult to design. Table Eighteen lists the disorders and current state of knowledge.

TABLE EIGHTEEN:
GENETIC FACTORS IN SPECIFIC DEVELOPMENTAL DISORDERS

Specific Developmental Disorder	Genetic factors
Developmental Arithmetic Disorder	Unclear if any familial factors.
Developmental Expressive Writing Disorder	No clear familial involvement.
Developmental Reading Disorder (Dyslexia)	Prevalence is estimated at 2-8 percent. This disorder is increased in other family members.
Developmental Articulation Disorder	In some form this disorder affects 5-10 percent of children. It occurs in a higher than expected incidence in other family members.
Developmental Expressive Language Disorder	It occurs in 3-10 percent of school age children. May be increased in family members.
Developmental Receptive Language Disorder	Prevalence 3-10 percent. No information on family data.
Developmental Coordination Disorder	Disorder of motor coordination. Prevalence 6 percent. No family info.

Since personality disorders are poorly defined and have low inter-rater reliability they are not good candidates for genetic analysis. Nevertheless, this is an area of intense interest and some very important work has been done. In general, it has long been suspected that constitutional factors played a role in personality disorders. Even the early psychoanalysts accepted some role for genetics in personality disorders. For many years it was felt that body type was related to personality. For example endomorphs (round body with tendency to baldness and obesity) were considered to have a temperament "which is said to be associated with love of comfort, sociability, extrovert interests, and a trend toward over relaxation, opportunism, or aggression (Kallmann, 1953)." However, poetic these findings are, they have contributed to a suspicion about the scientific basis of much of the personality research.

There have been a number of studies using twins to investigate normal personality variables, but much fewer studies of abnormal personality. In one study, Eyseneck and Prell (1951) investigated 25 twin pairs and determined that over 80 percent of "neuroticism" was due to genetic factors. Some studies have been done in specific personality disorders (Table Nineteen). Other personality disorders appear to be in the same "spectrum" of more common disorders. For example, in 1988, Squires-Wheeler et al. studied the rates of schizotypal personality disorder in 62 schizophrenic probands, 62 affective disorder probands, and 135 normal controls. Their results showed a schizotypal personality rate of 1.5 percent in the relatives of normal controls, and 7.9 percent in schizophrenia relatives (5x risk), and 11.3 percent in affective disorders (7.5x risk). The implication is that schizotypal personality disorder is an etiologically related variable phenotype that is in a spectrum with schizophrenia. One specific personality disorder that does appear to be highly familial is antisocial personality disorder, which has a prevalence of about 3 percent in males and 1 percent in females. It is five times more common in male family members and 10 times more common in female family members of affected probands. Also noted was increased somatization and substance use disorders in other family members. Adoption studies show an increased risk of antisocial personality in adopted out children.

TABLE NINETEEN:
GENETIC FACTORS IN SPECIFIC PERSONALITY DISORDERS

TYPE OF PERSONALITY DISORDER	EVIDENCE FOR GENETIC FACTORS IN CAUSATION
Paranoid	Undetermined
Schizoid	Undetermined
Schizotypal	Family and adoption studies
Antisocial	Familial and adoption studies
Borderline	Undetermined
Histrionic	Increased in family members
Narcissistic	Undetermined
Avoidant	Undetermined
Dependent	Undetermined
Obsessive Compulsive	Increased in family members
Passive Aggressive	Undetermined
Sadistic*	Undetermined
Self Defeating*	Undetermined

* Proposed personality disorders

Gottesman (1963) used the MMPI instrument combined with a twin study to demonstrate the high heritability of social introversion, psychopathic deviance, and schizoid reaction. Shields and Slater (1971) used the twin study technique to show a 29 percent MZ concordance and a 4 percent DZ concordance for a mixed group of personality and neurotic disorders. Inouye (1972) reported a 77 percent (27/35) MZ concordance and a 0 percent DZ concordance for obsessive-compulsive traits and a 21 percent (9/42) MZ concordance and a 0 percent (0/43) DZ concordance for hysteria.

CONCLUSIONS FROM THE STUDY OF THE GENETICS OF BROAD BEHAVIORAL PHENOTYPES

The genetic study of broad behavioral phenotypes represents the bulk of human behavioral genetics research. The results of these studies have yielded much useful information, but they have not added much to the critical subject of gene localization or mode of transmission. Table Twenty contains the current status of the role of genetics in the major broad behavioral phenotypes.

TABLE TWENTY:
OVERVIEW OF THE EVIDENCE SUPPORTING THE ROLE OF GENETICS IN BROAD BEHAVIORAL PHENOTYPES

Specific Disorder	Family	Twin	Adoption
Schizophrenia	X	X	X
Bipolar disorder	X	X	X
Major depression	X	X	?
Mental retardation	X	X	X
Alcoholism	X	?	X
Autism	X	X	
Tourette syndrome	X	X	
Alzheimer disease	X	X	
Stuttering	X	X	
Dyslexia	X		
Functional enuresis	X		
Panic disorder	X		

X = Conclusive evidence.
? = Some evidence, but not conclusive.

Chapter 7

Strategies to Localize the Genetic Factors in Broad Behavioral Phenotypes

To say that a broad behavioral phenotype has a strong genetic influence is, by itself, not very satisfying. What is really desired is to find the specific role that the genetic influence plays. How is the trait transmitted? What gene (or genes) causes the disorder? How does the defective gene result in abnormal behavior? These questions can only be answered if genes are localized for the broad behavioral phenotypes. There has been a massive search for the past twenty years to find a gene for a broad behavioral phenotype – a so-called "psychgene." So far there has not been the type of clear-cut success hoped for, however, powerful new tools have been developed that promise breakthroughs in the future. We will now review the strategies employed to find psychgenes.

THE ROLE OF ASSOCIATION STUDIES

Association studies are designed to measure the frequency of biological markers in individuals with a given disorder and to look for deviations in that frequency from expectations. For example, disease X has a prevalence of about one percent and 100 persons with the disease are tested to see if they have enzyme A or enzyme B. On testing 98 have enzyme A and 2 have enzyme B. Can it be concluded from this that enzyme A is associated with disease X? No. Even though there is a marked prevalence of individuals with disease X also with enzyme A, nothing can be concluded without some idea of the general population frequencies of enzyme A and enzyme B. To account for this, controls have to be used. This is the case-control method of association studies. The researcher repeats the study with a new group of 100 individuals with disease X and 100 individuals without disease X (matched for age). This time he finds 96 of the individuals with disease X have enzyme A and 4 have enzyme B. However, in the control group 70 have enzyme A and 30 enzyme B. This is not as convincing an association between disease X and enzyme A since enzyme A is very common in the control group also. However, the differences would appear to be deviations from chance (96 versus 70 expected, which may appear to signify an association). However, on further study the investigator realized that the case and control groups were not matched for an important variable – ethnicity. There were major differences in the ethnic makeup of the case and control group. This can be a problem because of potential differences in the prevalence of enzyme A and enzyme B in different ethnic groups. When the study was repeated with carefully matched ethnic groups the rate of enzyme A was 95 in the case group and 93 in the control. No association exists.

Association studies have a strong appeal to researchers because they are relatively simple and of reasonable expense. There are no shortages of patients or biological markers hypothesized to play a role in the various disorders. Also, to be considered a positive association, usually only a significant deviation defined as probability $< .05$ is needed. The negative aspects of association studies are much more troublesome. A major problem is the fact that even if a positive association is found, it simply does not imply any cause and effect relationship between the marker and the disorder. It essentially proves little, if anything. Secondly, the likelihood of false positives is very high because one in twenty association studies is likely to be positive due just to chance based on the accepted level of certainty. In addition, it is considered almost impossible to truly match cases and controls, which has led to repeated positive association studies, which after generating initial excitement, have been shown to be due to nothing more than differences between the ethnic backgrounds of the case and control groups.

In behavioral traits, HLA antigens have been the biological

marker of choice and have been tested ad nauseam. Additional choices include ABO types and other readily available polymorphic markers. More recently, there has been a strong tendency to use DNA polymorphisms in known genes, especially candidate genes (genes that can be postulated to play some role in the disorder). The thinking is that a positive association with a gene could imply a direct casual relationship when using candidate genes. A recent advance in the use of association studies is the introduction of the "haplotype relative risk" method of association study, which uses family members as controls and eliminates the ethnic background problem of the case-control method. This is now the clearly preferred method (Falk & Rubenstein, 1987, Terwilliger & Ott, 1992).

LOCALIZING GENES WITH CLINICAL HINTS FROM UNIQUE CASES

A second source of information that can lead to localization of genes is rare clinical cases that provide a hint about a possible relationship between a disease and a chromosomal region or gene. The two best examples of this are chromosomal defects associated with a known genetic disease and the clinical "linkage" of two diseases suggesting genetic proximity.

The Role of Unique Chromosomal Abnormalities

A strategy to potentially localize genes involves looking for clinical examples of chromosomal abnormalities, primarily deletions and translocations, that segregate with a disease phenotype in a family. For instance, if a family member had a disease and was shown to have a translocation involving chromosome 12 and if two other family members had the same disease and the same translocation, it would implicate the region of the translocation on chromosome 12. This approach is a long shot at best and there are only a few reported associations from this technique. Two prominent examples have been reported in behavioral traits. One involving a heteromorphism on chromosome 15 associated with the transmission of dyslexia. A second report of a chromosome 5 lesion segregating with schizophrenia has also been reported. However, neither of these reports has led to the defining of a specific gene.

Case Reports of Disease Associations

Another potential strategy to identify genomic regions of interest are reports of one disease presenting with another genetic disease. For instance, if a family was found where all family members who had one disease also had another disease, and the two diseases were usually not related, then the implication is that the two diseases may be genetically linked. Often these reports are not significant, because the chances of their mutual occurrence is the sum of their independent probabilities and the finding can be explained by chance alone. If a disease is found in relation to a rare disease in a much greater than statistically predicted pattern it may suggest linkage between the two disease entities. If the other disease has a known genetic location, this may lead to a gene localization. Again, this is a very rare occurrence given the vast number of genes, and when it does occur it is usually in isolated populations.

THE ROLE OF LINKAGE STUDIES IN LOCALIZING GENES

The most promising of the recent approaches to localizing genes is the linkage study paradigm. This concept of linkage first described by Morgan from 1910-1930 has slowly evolved to the point where it now is the major approach to gene localization. Localization of a gene is the first step in a process known as "reverse genetics." In this scenario, a gene is first localized by using linkage studies, then other techniques are used to "tighten" the linkage until the gene itself is located and cloned. Analysis of the gene product then reveals the specific defect, which completes the cycle. The importance of this process is that no knowledge of a disorder's biochemical defect is necessary to initiate these studies, a fact which is crucial to psychgenes given the limited amount known about their specific biochemical effects. With the isolation of a specific gene defect, this type of project would undoubtedly lead to new opportunities for improved diagnosis and treatment. An understanding of how this technique works is dependent upon knowledge of the concept of linkage and the role that DNA polymorphisms are playing in accelerating human linkage studies.

The Revolution in Gene Mapping Using Polymorphic DNA Markers

Advances in recombinant DNA techniques have sparked a revolution in gene mapping over the past decade and a half (White et al., 1987; White and Lalouel, 1988). Prior to the use of direct DNA markers, researchers had to use a relatively small number of biological markers such as blood groups and HLA, which severely restricted the chance of finding a linkage. In 1980, Botstein et al. outlined how DNA polymorphisms could be used to provide a new and rich source of genetic markers for linkage analysis. These markers are based upon differences in the sequence of DNA among individuals, and can be detected by restriction enzymes and Southern blotting techniques. Such markers, known as restriction fragment length polymorphisms (RFLP's), exist throughout the genome, even in areas that do not code for gene products. In addition, many RFLP's are highly polymorphic (Wyman and White, 1980; Jeffries et al., 1985; Nakamura et al.,1987), which greatly increases their usefulness as markers. Since the first arbitrary polymorphic DNA marker (a marker based on DNA differences without reference to a gene product or regulatory region) was developed in 1980 (Wyman and White, 1980), the number of markers and their distribution have increased almost exponentially. Arbitrary markers are combined with the large number of mapped genes to form "linkage maps" (White and Lalouel, 1988). Linkage studies using markers from these maps contributed to the localization of a number of important genetic diseases within the first decade of their availability (See Table One).

Linkage Basics

Simply stated, linkage studies are designed to statistically "link" the inheritance of a trait or disease (with unknown chromosomal location) to a marker that already has a documented chromosomal location. Geneticists have used linkage concepts to map the chromosomal locations of traits since 1913, when A. H. Sturtevant and Thomas Hunt Morgan clearly demonstrated that in the fruit fly Drosophila genes close together on the same chromosome are inherited in a pattern different from classical Mendelian independent assortment. What they observed was that genes close together on the same chromosome tend to be inherited as a unit, as if they were "linked" together. He also noted that as the distance between the genes increased they appeared to be less linked.

TABLE ONE:
GENETIC DISEASES MAPPED BY RFLP MARKERS

Year	Disease	Localized	Reference
1983	X-linked mental retardation syndrome	Xp27	Camarino et al.
1983	Huntington disease	4p16.3	Gusella et al.
1985	Cystic fibrosis	7q22.3-23.1	Wainwright
1985	Adult polycystic kidney disease	16p13.31-.12	Reeders et al.
1986	Chronic granulomatous disease	X p21.1	Baehner et al.
1986	Duchene and Becker muscular dystrophy	X p21.2	Monaco et al
1986	Retinoblastoma	13q14.1	Friend et al.
1987	Familial polyposis	5q	Leppert et al.
1987	Von Recklinghausen neurofibromatosis	17 paracent	Barker et al.
1988	Friedrich's ataxia	9p22cen	Chamberlain
1988	Benign familial neonatal convulsions	20	Leppert et al.

To understand this concept better, it is useful to evaluate the patterns of inheritance of two traits using several different types of relationships. For example, what would the inheritance pattern of two separate traits be if they were on different chromosomes? If a man has two distinct traits represented by the symbols "a" and "b," and he marries a woman who is "normal" for both traits (symbolically represented as "A" and "B"), the offspring will be affected as follows. One-fourth of his children will be unaffected (AABB), one-fourth will have "b" (AABb), one fourth will have "a" (AaBB) and one-fourth will have both (AaBb). Although the traits both existed in the father, each are represented in 50 percent of the children. This is what is meant by *independent assortment*. Each trait was inherited in a manner that did not effect the inheritance of the other. If the same cross is made, but with the disorders placed close together on the same chromosome, a different phenomenon called *non-independent assortment* occurs. In this case, the offspring will be of only two types. Half will be unaffected (AABB), and half will be affected by both disorders (AaBb). Compare this to the offspring ratio of independent assortment and notice the difference that the AABb and AaBB genotypes are not present. This is because "a" and "b" genes are linked (they are on the same chromosome), and they *assort* together, not independently. It became clear to early investigators, however, that this theoretical non-independent assortment did not occur to the extent predicted. For instance, the job of the geneticist would be easy if all genes on each of the 23 pairs of chromosomes in man were inherited in these linked packages, but as we know, this is not the case. The confounding factor is the process of *crossing over*.

Crossing over occurs when germ cells (egg and sperm) undergo meiosis to form gametes. Early in meiosis, each of the chromosome pairs replicates and then pair up, awaiting cell division. During this waiting period, each chromosome pair is represented by two copies, called chromotids. These chromotids can cross over each other and reform with a piece of the other chromosome. This process produces two additional gametes than the two already produced in non-independent assortment. These new gametes are called "recombinants" because they are formed from recombination of the two chromosomes. The four possible gametes are the same as those formed by independent assortment. A significant difference, however, is the ratios of the gametes. The best illustration of this principle is to take the same scenario as used above, and to assume that the location of the traits are unknown. If 50 couples with the male AaBb and 50 females with AABB had 100 children, the ratio of the possible offspring will predict the possibility of linkage. If all four phenotypes AABB, AABb, AaBB, AaBb are represented in a nearly 1:1:1:1 ratio, this suggests independent assortment (no linkage). If there is a preponderance of non-recombinant phenotypes (AABB, AaBb) versus recombinants (AABb, AaBB), linkage is suggested.

If two traits are on the same chromosome, the chance of crossing over to create recombinant gametes is increased as the physical distance between the two traits is increased. Because of this, two traits on different ends of a chromosome may not appear to be linked because the distance allows for many crossing over events to occur and recombinant gamete formations to take place. However, if two traits are very close, there may be virtually no recombinant gametes formed, because the chance of crossing over is very slim in such a narrow space. The comparison of the ratio of non-recombinants to recombinants is 1:1 in independent assortment. If the ratio is much higher than this (for example 2:1) there is evidence that linkage is present. The higher the ratio is the greater the likelihood of linkage.

There are two types of linkage studies in humans that are relevant to psychiatric genetics. The first is a *pedigree study*. In this type of study the inheritance of markers and traits are analyzed in families. Most psychiatric genetic linkage studies are of this type. The second type of study is a *half sib study.* In this type, sib-pairs (affected sibling pairs) are compared for the presence of the marker and trait in question. This technique uses the fact that siblings identical for the disease trait in question will likely inherit not only an identical disease gene, but also some portion of the surrounding DNA. For this reason, they are also likely to have common nearby markers. Such siblings are referred to as *identical by descent.*

Experimental Procedure of DNA Marker Based Linkage Studies

DNA purification and restriction endonuclease digestion - DNA is collected from white cells taken during phlebotomy. Some of the white cells are usually immortalized with a virus and kept in tissue culture as permanent source of DNA. When the transformed "immortal" cells are growing in culture, small amounts can be removed, and the DNA extracted. Once the DNA is purified it is present in long strands of intermingled microscopic fibers. Restriction endonucleases are enzymes used to cut the DNA. They cut at specific DNA sequences only. For example, "Eco R1" from

E. *coli* cuts at the sequence GAATTC, but not any other sequence. As a result, the DNA is fragmented in specific reproducible pieces. However, it is also known that differences exist between individuals, and that some people have more or less recognition sites than others. This, of course, is the basis of the polymorphic nature of these DNA pieces.

Electrophoresis of fragmented DNA and the gel pattern transfer - The fragmented DNA is then electrophoresed. After electrophoresis, thousands of different size fragments form a spectrum across the gel with the smallest fragments moving farthest, and the largest moving the least. The next step is to make a template of the gel by blotting it with a nylon filter, which transfers DNA fragments to the filter in an exact template of the original gel. This system is known as the "Southern blot." This transfer of DNA to the nylon filter allows for multiple uses of the same blot.

Probes and hybridization - The next step is to hybridize a probe to its area of homology on the spread of DNA fragments. Each probe is incubated with a spread of DNA and then the blot is washed. Only the probes that have found an area of homology are not washed off at this point. The next step is to hybridize the DNA probe with the DNA coated paper. Since the probes are radiolabeled, the paper is sandwiched with a negative. The developed plate shows the image of the probe in relation to the gel. The key to RFLP linkage studies involves the proper use of probes. A probe is a piece of DNA that hybridizes to the corresponding sequence of DNA in the cell. They are made from either reverse transcription of mRNA, known as cDNA, or they can be sequenced to correspond to known protein structure. There are two sources of probes that can be used; probes known to hybridize with RFLP's in the general population, and probes that will hybridize with single copy genes (this includes certain candidate genes that might play a role in behavioral disorders).

Statistical Analysis of Linkage Data

Analysis is done based on the concept of lod scores, which is a statistical way to measure the probability of linkage. The higher a lod score is, the more probable that linkage is occurring. If a lod score is 3.0 for instance it indicates that such a distribution could occur by chance only 1 in a 1,000 times. Lod scores work for large individual families as well as for the addition of several smaller families (in the absence of heterogeneity). However, the more families tested the more predictive the lod score becomes. Sophisticated computer models of linkage are available that determine a lod score. Morton described the modern lod score in 1955. By 1974, Ott had developed the LIPED program to determine lod scores by computer. Lathrop improved this technique in 1985 with multipoint linkage analysis.

Strategies for Linkage Studies

One approach to detect linkage is to methodically test the human genome using evenly spaced polymorphic markers from the human linkage map. The number of markers necessary for this approach can be roughly estimated as follows. Geneticists use the term *centimorgan* to define a distance on a chromosome. Specifically, it is the distance between two chromosomal regions where crossing over occurs in only one of a hundred meiosis. In physical terms, this is generally accepted to be approximately one million DNA base pairs. Since there are approximately 3.3 billion DNA bases in the human genome (one set of 22 autosomal chromosomal pairs and one sex chromosome pair) it follows that roughly 3,300 centimorgans in the human genome (Renwick, 1971). This implies that recombination will occur on average only 1 time in 10 meiosis in a span of DNA comprising 10 centimorgans, or, alternatively, ninety percent of such crosses would not produce recombination. The recombination rate in a span of ten centimorgans is, therefore, a rare enough occurrence that linkage will be readily detected to any marker within this distance. If markers were chosen approximately 20 centimorgans apart, it follows that any causative gene would have to be located within 10 centimorgans either side of the marker. This suggests that 165 evenly spaced markers (3300 centimorgans divided by 20 centimorgan blocks) could provide enough polymorphisms to saturate the human genome for linkage study purposes (Botstein et al., 1980). Unfortunately, markers are often not sufficiently informative (they do not provide any information useful to test for linkage), and equal spacing of markers can be difficult to establish so that the actual number of markers needed must be increased (Lange and Boehnke, 1982). However, even if the number needed is doubled, it would still take less than 400 markers to methodically test for linkage throughout the human genome.

Another approach would be to select (or avoid) markers from the human linkage map based upon an *a priori* assignment of high or low linkage likelihood. The determination of the linkage likelihood of a specific marker is based upon a synthesis of data from evidence concerning the mode of transmission, previous linkage studies, chromosomal abnormalities, case reports and known locations of "candidate genes" for psychgenes. The potential advantage of this technique is that it may lead to a finding of linkage with fewer markers being used than in the methodical approach.

The Consequences of Establishing Linkage

The goal of performing linkage studies in psychiatric disorders is to unambiguously link a chromosomal region to a broad behavioral phenotype. If such a linkage is established, a number of important conclusions and research opportunities will arise. The most immediate conclusion will be proof that at least one form of the particular disorder is caused by a genetic defect. After an unambiguous linkage is achieved, the next step will be to determine the level of heterogeneity. By performing linkage studies on a large number of families some indications of the extent of heterogeneity will be obtained. Families that do not show linkage will be tested further and other genetic forms of the disease may be found. Even with heterogeneity, several important projects can be started with an unambiguous linkage. The first goal would be to use the linkage as a means to narrow down the search for the actual genetic defect. This most likely will mean attempting to clone the defective gene (although defects in regulatory regions might also be the defect). Finding and cloning a gene, even with linkage to a DNA probe, is difficult. Even with linkage, the actual gene may be up to a few million base pairs away. So it then becomes a problem of "tightening" the linkage until the actual gene is obtained. Tightening is done by trying to find closer markers to the gene, by "walking" the chromosome, or by using non-methylated regions to seek out likely coding regions nearby. The biggest obstacle to finding the actual gene is the large number of genes in any area of the human genome. Estimates put the total gene number at 50,000 to 100,000, of which roughly 30,000 may be "brain" genes. They are spread out over approximately 3.3 billion base pairs. So, even if an area of 1 million base pairs is found, there could be up to 25 genes, 10 of which are genes involved in the structure or function of the brain.

The immediate benefit of cloning the gene would be to determine the function of the protein. Studies may be able to show how the gene defect either makes the protein unusable, or how the defect changes its ability to function. In addition, once a psychgene is cloned it can be sequenced and compared to its normal counterpart. By selecting restriction endonucleases so that the defect would be identifiable by a RFLP, the sequenced gene could provide an extremely accurate method of diagnosis. A diagnostic test based on this could be used as a mechanism to detect asymptomatic carriers very early, and could even be used prenatally, although that use brings up many difficult ethical issues. If the actual protein defect was known, specific pharmacologic therapies could be matched to the disease process. The benefit to affected individuals could be enormous. There are no trait dependent markers for broad behavioral phenotypes. If linkage could be established this would change. The presence of a "marker" for a psychgene would allow for novel studies into the role of the environment in these disorders. It would be possible to identify those patients that carry the psychgene *before* symptoms appear. This would allow for experimental analysis of the role of environment in these diseases. For instance, a cohort of paired family members one with the gene and one without could be followed *prospectively* with careful attention to environmental issues. Some of these environmental factors may emerge as contributory to the development of the illness. Those environmental factors shown to be important could then be avoided in known high risk groups. It may be that no environmental factor will be identified. In short, the results of these types of studies could answer some of the most basic questions.

Potential Limitations of Linkage Studies

Counterbalancing the promise of linkage studies is a number of significant challenges. The first and potentially most limiting is the possibility that the genetic component of broad behavioral disorders is not consistent enough to be detected by linkage techniques. Some forms of polygenic inheritance where multiple genes combine to exceed a threshold to cause a phenotype are considered poor candidates for linkage studies. Supporting this hypothesis are results of computerized analyses of various family pedigree data, which in general are more consistent with polygenic models. Even though several loci might determine the expression of a psychiatric disorder, it is still possible that a single loci can play a significant role in the etiology of these illnesses. Genotypes must be compared to phenotypes for linkage studies to work. Accurate genotypes are a result of specific biochemical analyses of donor DNA. They are specific, reproducible, and have very high reliability. In contrast, the equally important phenotypes are based only on a gold standard of the diagnostic interview.

The Future of Linkage Studies in Psychiatric Genetics

The future of linkage studies in psychiatric research is very promising. The tools necessary to perform linkage studies are improving rapidly. At the current rate of new marker determination, a high density linkage map of the entire human genome is in the very near future. The use of polymerase chain reaction (PCR) based markers has greatly accelerated this progress. This implies that any disease phenotype that is determined by a single locus genotypic alteration can eventually be localized. The quality of individual markers is also improving. New markers are increasingly polymorphic. Multipoint mapping is also improving the localization of genes by using data from several markers at once.

Chapter 8

The Current Status of Attempts to Localize Genes for the Broad Behavioral Phenotypes

THE SEARCH TO LOCALIZE A GENE FOR SCHIZOPHRENIA

Association Studies in Schizophrenia

McGuffin et. al. (1986) reviewed the schizophrenia association data and found no conclusive findings. Positive association does not generally imply linkage, however, given founder effects, some populations could show evidence suggesting linkage from a strongly positive association study. This happens in genetic isolates, because too few generations have occurred for random recombination between the association site and the deviant gene to have taken place. Prominent gene product polymorphisms studied for association include Gc (4q12), HLA (6p21.3), ABO (9q34), C3 (19p13). The most prominent associations are HLA (6p21.3) subtype A9 and paranoid schizophrenia and C3 (19p13) with schizophrenia (Rudduck et al., 1985).

Unique Clinical Cases with Hints about Localization

Two prominent reports in the literature describe cytogenetic abnormalities that are present with the schizophrenia phenotype. Bassett et al. (1988) reported a chromosome 5 trisomy (5q11.2 to 13.3) cosegregating with schizophrenia in one family. Chodirker et al. (1987) report the presence of a folate sensitive autosomal fragile site on 19p13 in four brothers. Two had schizophrenia and one had autism. These findings may be chance occurrences, or they may signify that a genomic region associated with schizophrenia is located within or near these regions. Baron (1976) reported the presence of a family with oculocutaneous albinism (Xp22) that also was segregating a schizophrenia-like psychosis (lod score of 1.55). Eye color (polygenic) (Constantinidis, 1958), and low dopamine beta hydroxylase levels (9q34) (Book, 1978) have been speculated to be associated with the schizophrenia phenotype. None of these associations have been conclusively proven.

Candidate Genes

Schizophrenia "candidate" genes are genes that code for proteins that have a theoretical role in the pathophysiology of schizophrenia. The chromosomal locations of many candidate genes are known (Gurling, 1986). For instance, tyrosine hydroxylase is the rate limiting step of dopamine metabolism (Lamaroux et al., 1982), and is located near the tip of the short arm of chromosome 11 (Craig et al., 1986). The selection of a polymorphic marker, such as H-Ras, that is only a small genetic distance from the tyrosine hydroxylase gene insures that linkage will be detected if it is present. Candidate genes can be confirmed or ruled out very quickly using this approach. A problem with this strategy is that as many as one third of the 100,000 or so human genes may be involved in brain structure and function, so finding linkage near a candidate gene still does not prove that it is the pathologic gene. For example, if linkage as close as one centimorgan (one million base pairs) is obtained, as many as nine "brain genes" may be found within this limited area (30,000 genes / 3300 centimorgans). Another possible strategy is to use the cloned gene itself as a marker for linkage. With this approach an etiologic role can be ruled out by any number of recombinants, since no chance of crossing over is expected. Choosing candidate genes may be a very difficult task. Many interesting loci will prove to be false leads, as should be expected given that any one of the 30,000 genes may be involved. Dopamine-ß-hydroxylase, for instance, is a gene that has been implicated in schizophrenia (Book, 1978), and could be considered a candidate gene. The recent finding that it is linked to ABO (9q34)

(Wilson, 1988) makes it an unlikely factor in schizophrenia because the blood group loci has been excluded for linkage in two studies (McGuffin et al., 1983; Andrew et al., 1987). In addition to the enzymes, peptides, and receptors, other types of brain proteins exist, including ion channel proteins, structural proteins of axons, dendrites, and synapses, and proteins (DNA binding proteins) that regulate transcription or translation of a brain peptide or protein. The genetic alteration may not even involve a protein, as it is possible that changes in regulatory DNA sequences may be responsible for schizophrenia. With this abundance of potential candidate genes, it becomes increasingly difficult to differentiate between interesting candidate genes, further limiting their usefulness. In addition, many candidate genes that have been cloned are not useful markers because of their lack of polymorphism.

Linkage Studies in Schizophrenia

The goal of linkage studies in schizophrenia is to unambiguously localize the expression of the schizophrenia phenotype to at least one chromosomal location. There are a number of advantages to using linkage studies in schizophrenia. First, the level on inquiry is at the genotypic level, which is the clearest level of information. If schizophrenia is a genetic disease, the more removed the level of inquiry is from the genotype, the more confounding factors may be present. Second, linkage is not dependent upon identifying any specific biochemical defect in schizophrenia. In this approach the exact mode of transmission does not have to be known.

Linkage studies in schizophrenia involve significant difficulties, which may explain why the first linkage study was not attempted until 1973. Elston et al. (1973) performed a sib-pair analysis of previously collected twin data. As the authors admitted the sample size was small and no conclusive evidence for or against linkage was found to any of the eight polymorphisms analyzed. They did find evidence that suggested linkage to Gm (14q32.33), Rhesus (1p36-34), and Gc (4q12). The next study was done by Turner (1979) who presented data that suggested linkage with HLA (6p21.3). His positive finding of a lod of 2.57 to HLA has not been reproduced, and four subsequent reports found evidence for exclusion (McGuffin et al., 1983; Chadda et al., 1986; Goldin et al., 1987; Andrew et al., 1987). Four years later, McGuffin et al (1983) studied linkage between 19 markers and HLA. They showed evidence for exclusion at the PGM1 (1p22.1), Km (2p12), MNS (4q28-31), HLA (6p21.3), GPT (8q24), ABO (9q34), Pi (14q32.1), and ADA (20q13.11) locuses. An additional study by Andrew et al. (1987) presented exclusion data to HLA (6p21.3), Rh (1p36-34), and Gm (14q32.33).

Four studies have been reported using RFLP techniques. These reports are primarily focused on two areas implicated from recent reports. The first is a trisomy on the long arm of chromosome 5 (5q11.2 to 13.3) that has been shown to segregate with schizophrenia in one family (Bassett et al., 1988), and the second is the finding that the tyrosine hydroxylase gene is located near the tip of the short arm of chromosome 11 (Craig et al., 1986). Byerley et al. (1988) presented data from three families that does not support linkage in either of these areas. Gurling et al (1988) reported a lod score of 1.0 at a recombination fraction of 0.2 for the glucocorticoid receptor that is located in proximity to the 5q trisomy, and later reported linkage in this area with two probes (Sherrington et al, 1988). St Clair (1988) reported no evidence for linkage using the Harvey-ras-1 and insulin probes to test 11p.

Of the linkage studies in schizophrenia only the 5q finding by Sherrington et al. shows definite linkage (1988). (See Table One). However, this study has never been replicated. Evidence for exclusion of linkage is presented at eleven loci based upon single reports. If lod scores for a loci are summated from the different studies, the number goes up to thirteen. It is apparent that because of heterogeneity and the small fraction of the genome tested, exclusion of any genomic region may be premature.

In 1994, Coon, et al. reported on a massive screening of the human genome for schizophrenia loci. Area 22q11 was identified as an area of interest. Pulver, et al. subsequently reported additional evidence for this area.

TABLE ONE:
PREVIOUS LINKAGE STUDIES IN SCHIZOPHRENIA

#	Year	Author	Findings
1)	1973	Elston, et al.*	Suggests linkage to Gm, Rhesus, and Gc**
2)	1979	Turner ***	Suggests linkage to HLA
3)	1983	McGuffin, et al.	Excludes linkage with HLA & 6% of genome
4)	1986	Chadda, et al.	Excludes linkage to HLA
5)	1987	Goldin, et al.	Excludes linkage to HLA
6)	1987	Andrew, et al.	Excludes linkage to HLA, Rh, Gm
7)	1988	Byerley, et al.	Preliminary data does not suggest linkage to areas of 5q and 11p in three families
8)	1988	DeLisi, et al.	Suggests 5q linkage
9)	1988	Gurling, et al.	Evidence suggestive of linkage at 5q
10)	1988	St Clair, et al.	No evidence for linkage to 11p
11)	1988	Sherrington, et al.	Claim of linkage to 5q in seven families
12)	1988	Kennedy, et al.	No evidence for linkage on 5q in one family
13)	1994	Coon, et al.	Suggest area 22q11.
14)	1994	Pulver, et al.	Evidence for area 22q11.

* Sib pair study

** Gc linkage associated with broader diagnosis criteria including schizophrenia spectrum disorders.

** Diagnosis based on "schizotaxia" which includes schizophrenia spectrum disorders.

Conclusion

Linkage analysis using RFLP markers is likely to be a valuable tool in understanding schizophrenia. The development of strategies for linkage studies may potentially reduce the time and costs involved in confirming current markers and finding new ones. Clear establishment of linkage to one or more markers could have many consequences. Close linkage of schizophrenia to a polymorphism could be an effective trait dependent marker leading to an improvement in diagnosis. Another use for the marker would be to identify a cohort of pre-symptomatic affected individuals and follow them to analyze environmental determinants that may predispose or protect them from developing the schizophrenia phenotype. A gene linked to schizophrenia could eventually be cloned, which should contribute dramatically to our understanding of the pathogenesis of the disorder. Even if a localized gene represented only a small percentage of the cases of schizophrenia, valuable insights into the pathophysiology of the other forms may be possible. For example, Brown and Goldstein's (1985) study of rare forms of familial hypercholesterolemia has had tremendous impact on the understanding of all forms of hypercholesterolemia. In summary, given the power of the linkage techniques now available, and the potential benefits, the interest in these studies for schizophrenia will only increase in the future.

CURRENT STATUS OF LOCALIZING GENES IN BIPOLAR DISORDER

Association Studies

MAO gene - Lim et al. (1994) reported a positive association between monoamine oxidase A gene and bipolar affective disorder using a dinucleotide repeat polymorphism (n = 57). The study was done using populations of Western European extraction. Kawada et al. (1995) repeated this finding using the same polymorphism in a different population (Japanese)(n = 58). Both of these studies used the classic case-control design that is prone to error due to differences between the study group and controls, no matter how much care is used in the matching. Nöthen et al. used the preferred haplotype relative risk method and was unable to show any evidence for an association between MOA A gene and bipolar disorder (n = 82). It appears that this association is another false positive due to small sample size and to subtle differences between the study and control group. Direct sequencing of the MOA gene would provide the definitive answer.

Linkage Studies in Bipolar Disorder

Family, twin and adoption studies all implicate a significant genetic factor in the etiology of bipolar disorder. However, the exact molecular and biochemical mechanisms involved have not been identified. If this clinically heterogeneous group of disorders is caused by genetic defects, the isolation of the responsible regulatory region or gene product associated with the bipolar disorder phenotype would be considered proof of the genetic etiology. The first step in that direction is to "link" the phenotypic expression of bipolar disorder to a known chromosomal location. Since the discovery of restriction fragment length polymorphisms (RFLP's), there has been a marked increase in the interest in linkage studies for bipolar disorder. This new technology has revitalized linkage studies by providing a source of abundant polymorphic markers. It would be a mistake, however, to see the use of DNA markers as the start of linkage studies in bipolar disorder, because around thirty linkage studies have already been reported in the bipolar disorder literature (See Table Two).

TABLE TWO:
LINKAGE STUDIES IN BIPOLAR DISORDER

#	Year	Author	Type of Study	Findings
1)	1968	Tanna and Winokur	Sib Pair	No evidence for linkage to ABO Blood Group
2)	1969a	Winokur and Tanna	Pedigree	Findings suggest that further studies should be done testing X- linkage.
3)	1969	a) Reich, et al. b) Winokur, et al.	Pedigree	First claim of proof for X- linkage to color blindness loci.
4)	1972	Mendlewicz, et al.	Pedigree	Claim of linkage to deutan and protan color blindness on the X chromosome.
	1973	Fieve, et al.	Pedigree	Suggestive Xg linkage in four families.
	1975	a) Mendlewicz	Pedigree	
		b) Fieve, et al.	Pedigree	Preliminary report of study published in 1974.
	1974	Mendlewicz, et al.	Pedigree	Claim of linkage to protan and deutan color blindness and Xg. Summation of all previous studies.

TABLE TWO: (CONT)

5)	1976	Belmaker, et al.	Pedigree	Minimal evidence of Xg linkage in one family.
6)	1977	Baron, et al.	Pedigree	Suggestive evidence for linkage with color blindness.
7)	1977	Johnson, et al.	Pedigree	No evidence of linkage to color blindness.
8)	1978	Smeraldi, et al.	Sib-pair	Possible linkage to HLA.
9)	1979	Targum, et al.	Pedigree	Not linked to HLA.
10)	1979	Gershon, et al.	Pedigree	No linkage to color blindness found in six famililies.
11)	1979	Leckman, et al.	Pedigree	No linkage to Xg.
12)	1979	Mendlewicz, et al.	Classic	X-Linkage to color blindness found to be heterogeneous in eight families.
13)	1980	Gershon, et al.	Pedigree	Collaborative study. Combination of Mendlewicz and Gershon 1979 studies with the addition of two new families.
14)	1980	Mendlewicz, et al.	Pedigree	Claim of linkage to G6PD.
15)	1981	Smeraldi & Bellodi	Sib-pair	Suggest linkage to HLA.
16)	1981	Weitkamp, et al.	Sib-pair	Suggest linkage to HLA.
17)	1982	Kruger, et al.	Pedigree	Suggest evidence of linkage of depressive "taxon" to HLA.
18)	1982	Goldin, et al.	Sib-pair Pedigree	No evidence of linkage to HLA.
19)	1982	Suarez & Croughan	Sib-pair	No evidence of linkage to HLA.
20)	1984	Del Zompo, et al.	Pedigree	Suggest evedence for linkage to CB-G6PD.
21)	1984	Kidd, et al.	Pedigree	Preliminary study in the Amish.
22)	1984	Gerhard, et al.	Pedigree	RFLP. Preliminary suggestion of linkage to 11p in the Amish.
23)	1987	Egeland, et al.	Pedigree	RFLP. Linkage to 11p in Amish.
24)	1987	Detera-Wadleigh	Pedigree	RFLP. No evidence for linkage to 11p in three North American families.

TABLE TWO: (CONT)

25) 1987	Hodgkinson, et al.	Pedigree	RFLP. No evidence of linkage to 11p in three Icelandic families
26) 1987	Baron, et al.	Pedigree	Claim of linkage to color blindness and G6PD.
27) 1987	Mendlewicz, et al.	Pedigree	RFLP. Claim of linkage of Taq1 polymorphism at Xq27.
28) 1988	Gill, et al.	Pedigree	RFLP. No linkage in one Irish family to 11p.
29) 1988	Waters, et al.	Pedigree	No evidence for linkage.
30) 1989	Berrettini, et al.	Pedigree	RFLP. No evidence for X- linkage. Rules out significant heterogeneity.
31) 1989	Kelsoe, et al.	Pedigree	RFLP. No chromosome 11 linkage in Amish Family.
32) 1993	Baron, et al.	Pedigree	RFLP. Lack of X-linkage.
33) 1994	Berrettini, et al.	Pedigree	RFLP. Claim of linkage to chromosome 18.
34) 1995	Pauls, et al.	Pedigree	RFLP. No linkage of Bipolar to 18.
35) 1995	Stine, et al.	Pedigree	RFLP. Further evidence for chromosome 18.

The first linkage study of a psychiatric illness involved bipolar disorder. Tanna and Winokur in 1968 used the sib pair method to test the ABO locus for linkage to "primary depression or mania." The study was limited by the use of only one marker (ABO) and the unclear distinction between unipolar and bipolar disorder. Nevertheless, it was an important contribution because it demonstrated the feasibility of linkage studies in psychiatric disorders.

A second study by Winokur and Tanna followed in 1969. They performed a pedigree linkage study using six families that had at least one member with bipolar disorder. They used ten genetic markers of which only 6 were informative in at least one family. Phase (whether the marker and the phenotype were coupled or in repulsion) was not determined except for Xg, but the ratios of recombinants to non-recombinants were provided. The final data on ABO, Rh, Duffy, MNS, Kidd, and Xg markers were limited in quantity, and the only significant findings were ratios of "recombinant to non-recombinant" of 1:7 between X-chromosome marker Xg and bipolar illness, and 1:10 for X-linkage in general. (This corresponds to maximum lod scores of 1.09, and 1.78 respectively) As noted by the authors "the data make X-linkage appear to be a reasonable hypothesis in manic-depressive disease . . . [so] . . . it is necessary to collect additional families . . ." It is important to realize that they did not see this as proof of linkage.

The third study was done virtually simultaneously by the same center and was published in one main article and one brief follow-up. Two families segregating for bipolar disorder and color blindness were studied. The statistical analysis was based upon the assumption that the probability of color blindness and bipolar disorder being present in the same individual was equal to $(1/2)^n$. All twelve cases carried both, so the odds favoring linkage were $(1/2)^{12}$ ($p < .0005$) that the two disorders were linked. This is not considered a useful way to analyze data by modern standards. Reanalysis can be done by studying the ratio of non-recombinants to recombinants in both pedigrees. Family one shows 3:0 non-recombinants to recombinants and family two shows 6:0 non-recombinants to recombinants. If both families are added then there are 9:0 non-recombinants to recombinants. Without knowing phase, this corresponds to a lod score of 2.37. This value is not generally considered proof of linkage, but it is suggestive. So, after three studies, there was no conclusive evidence for linkage, although the X-linkage theory was gaining momentum.

Three years passed until Mendlewicz et al. began a series of reports implicating X-linkage for bipolar disorder. They studied seven additional families for linkage to color blindness(CB) and found a ratio of 13:4 non-recombinants to recombinants. This corresponds to a lod of 1.06 assuming phase is known. If this total is added to the earlier data, then there are 22 non-recombinants to 4 recombinants or a calculated lod score of 2.97. The next study by Fieve et al (1973) reported data on linkage in four families to the X-chromosome marker Xg (lod of 0.69). These relatively low values are somewhat in contrast to the assertion that "they provide evidence for measurable linkage between the Xg blood group and manic-depressive illness." In fact, even after these five studies, neither Xg or color blindness (CB) had been shown to be statistically linked to the bipolar phenotype. Although, the CB data was very close to the 3.0 lod score necessary to show linkage. In 1974, the Mendlewicz group published the results of their ongoing linkage studies for CB (both the protan and deutan forms) and Xg. In summary, they found that in 10 families with deuteranopia and 16 with protanopia (including earlier families) a combined lod of 8.13 at a recombination fraction of .10. Their Xg linkage data in 25 families showed a lod score of 2.95. Thus, there appeared to be one positive study that showed that bipolar disorder was linked to the CB marker, but the Xg lod score was not at the value necessary for linkage.

In 1976, Gershon and Bunney critically reviewed the Mendlewicz et al. data, and they found several points of disagreement. First, they questioned how bipolar disorder could be linked to both Xg and CB when the markers are far apart on the X chromosome. Secondly, they questioned whether the color blindness data was a true case of linkage or a case of association. Third, they felt that the analytic method used in the study was defective because it failed to account for decreased penetrance and variable age of onset, and lastly, they felt that some of the pedigrees were open to other analyses than the one offered. It was clear at this point that the linkage of bipolar disorder to CB was not universally accepted.

In the next few years there were four new reports concerning X- chromosome linkage studies. Belmaker et al (1976) reported on a family of ten siblings with diagnoses including bipolar disorder, psychotic depression, acute schizophrenia, and schizoaffective illness. They showed evidence that non recombination occurred in 7 of nine siblings (maximum lod of 0.63) to the Xg locus. This report appears flawed because of the overinclusiveness of phenotypes associated with bipolar disorder, and in addition its lod score is far below any level of significance. A report by Baron (1977) suggested linkage (max lod of 2.33) of bipolar disorder to deutan color blindness in one family that had been identified by a proband with schizoaffective disorder. The controversial inclusion of schizoaffective disorder as a phenotype was later reviewed by Baron et. al., with the conclusion that it could be a valid phenotype based on the frequent occurrence of individuals with schizoaffective disorder in the pedigrees from previous linkage studies. This conclusion is still controversial and most investigators have tried to avoid the inclusion of schizoaffective disorder or schizophrenia in their bipolar pedigrees. However, this report was seen as further evidence for linkage of bipolar to CB.

The X-linkage findings became even more controversial in 1977 when Johnson et al. reported the first negative linkage study to color blindness (lod of -1.46). In 1979, Leckman et al. undertook a linkage study of Xg in six families and found no evidence of linkage (lod of <-2.0). So, by 1980 the Xg findings, which were never statistically strong, had lost their support altogether, and the CB findings were facing increasing challenges. In order to help clarify the situation, a study was begun by members of the World Health Organization Biological Psychiatry Collaborative Program. These investigators, which included all of the people most active in the field, presented data on linkage to color blindness in 16 families from four different centers. The results were, overall, not supportive of linkage to CB, but interesting hints of heterogeneity were present. Of the 16 families, 8 had positive lod scores, and one was 2.5. All the scores combined showed a maximum lod score of 1.0. The finding of heterogeneity was mostly due to significant differences between the six families ascertained in Bethesda by Gershon's group and the eight ascertained in Brussels by Mendlewicz's group. Gershon's data on six families showed a fairly homogeneous (for non-linkage) set with a lod < -2.0, while Mendlewicz's data showed a maximum lod of 1.55. The two independent families had a maximum lod score of 1.91. Mendlewicz et al. felt that the only explanation of the difference was heterogeneity, and pointed to evidence of heterogeneity within his own group of eight families. With the controversy growing, Mendlewicz et al. (1980) reported a linkage study of bipolar disorder with G6PD in one very large family. G6PD is genetically very close to the color blindness locus, so this was presented as further evidence of linkage of bipolar disorder to a locus near Xq27-28. The lod score generated was 4.04, well above the 3.0 necessary to claim linkage.

The X-linkage inconsistencies were addressed by Gershon in 1980, when he openly questioned several discrepancies. First he pointed out that color blindness and Xg loci could not be closely linked because they were far apart on the X-chromosome. Thus, implying that if the Xg data was incorrect maybe some of the CB data was flawed also. Secondly, he states that the strong linkage findings and homogeneity of Mendlewicz's early bipolar data were inconsistent with later findings and he suggests that "a methodological error" may have occurred:

> I urge future studies of linkage in psychiatric disorders to take great pains to document ascertainment and diagnostic procedures to demonstrate reproducibility of diagnosis and to take care that data from interviews, family histories, and medical records be retained for later review . . . Until then, it remains for me an open question whether linkage ever exists between bipolar illness and the protan-deutan-G6PD region of the X chromosome.

The X-linkage dispute was then reviewed by a Baron et al. (1981). They analyzed the previously published linkage data for errors of ascertainment and data analysis, but found little evidence to support any major methodological error that could cause false positive results. They agreed with Mendlewicz and concluded that bipolar disorder was a genetically heterogeneous disorder. In 1982, Risch and Baron reviewed the previous X -linkage data, correcting for age of onset and testing for heterogeneity. They concluded that the reports of linkage to the G6PD-color blindness area were supported and that reports of non-linkage were likely due to heterogeneity. The Xg findings were not supported, and they suggested that "future linkage studies be directed toward the G6PD - color blindness region on the X chromosome." Only one further X-linkage study was done over the next few years. A report of two families, one segregating color blindness and bipolar disorder, and the other G6PD and bipolar disorder. The two families combined produced only a lod score of 1.47.

The X-linkage question resurfaced again in 1987 when Baron et al. confirmed his earlier X-linkage work in a report on G6PD and color blindness (CB). Admitting heterogeneity and excluding all families with male to male transmission, he again found evidence for linkage to G6PD and CB. Four families were tested against CB, three showed linkage one did not. Using multipoint analysis he concludes "the most probable location of the disease locus is close to and possibly between the two marker loci." In conclusion, he speculated that "one third of the bipolar population carry the X-linked gene." In the same year, Mendlewicz et al showed evidence supporting linkage of bipolar disorder to coagulation factor IX at Xq27. They used a well known factor IX RFLP to study 10 bipolar pedigrees with no male to male transmission. The results showed a maximum lod score of 3.10 at a recombination fraction of 0.11 with full penetrance. These findings were offered as further proof of the terminal region of the long arm of chromosome X being involved in at least some forms of bipolar disorder.

At this point, the X-linkage controversy had reached a stalemate. Only Xg had been ruled out as a linked marker. In spite of the methodological questions, evidence was mounting for linkage of the three linked X-chromosome markers (CB, G6PD, and Coagulation factor IX) to the bipolar phenotype. However, these findings are again being questioned because of the recent study of Berrettini et al (1989) which excluded linkage among 10 pedigrees for the color blindness region using the flanking marker DXS52 (a lod of -8.27 was obtained). The study used only pedigrees lacking male to male transmission. They also addressed the question of heterogeneity by statistically determining that at a recombination fraction of .01 linkage could be ruled out for any percentage greater than 25% of the families tested. The authors considered this evidence against significant heterogeneity.

In 1978, a report by Smeraldi et al. began a controversy about possible linkage of bipolar disorder to the HLA system on chromosome 6. The HLA area was analyzed because of a number of association studies showed possible involvement of the highly polymorphic HLA system in the susceptibility to "primary affective disorders." Smeraldi et al. wondered if the weak association may be due to a "susceptibility" locus linked to HLA, so they undertook a sib-pair linkage study. The study showed a slightly statistically significant increase in the similarity of the HLA haplotypes between affected sib pairs. A classic pedigree study was reported shortly after this report by Targum et al. (1979). They found no evidence for linkage in nine families. In 1981, Weitkamp et al. published a modified sib pair study based upon a model used in juvenile onset diabetes that purported to show a genetic influence on affective disorders by a locus linked to HLA. Kruger et al (1982) also published an HLA linkage study based on two large pedigrees segregating affective disorders. The study reached a tentative conclusion of linkage with HLA (lod of 3.91 at a recombination fraction of 0.10), but they admitted the study was "based on several untested assumptions, including penetrance function and diagnostic spectrum and . . . therefore one cannot interpret the lod score statistics using the ordinary criterion of lod score . . ." The study was also unusual because the diagnosis of "other affective disorder" included dysthymia, borderline personality disorder, schizotypal personality disorder, obsessive compulsive personality disorder and childhood psychosis. Two groups reviewed the Weitkamp et al. data and found a methodological error. They showed that assumptions about arbitrarily dividing the sib pairs by Weitkamp were unfounded and that the linkage in his study was due to "different methods of data analysis" and not clear linkage. In addition, they published their own data, which did not show linkage to HLA. Goldin et al concluded that no linkage was found "whether by calculating the IBD distribution in affected sib-pairs, calculating lod scores under both a dominant and recessive model, or by dividing our families according to the criteria of Weitkamp." Interest in further pursuit of HLA linkage studies has waned.

In 1984, Kidd et al. published the first set of data on linkage studies from a genetically isolated Amish population in Pennsylvania. The Amish are considered ideal for research in the genetics of psychiatric disorders because:

> . . . the population is highly self-defined and closed, with little in-migration since the original progenitors. Complete demographic study has been done on the group and extensive genealogies were prepared. [They] have large sibships spanning three to four generations . . . There were no problems of uncertain paternity. Since few people leave the community, it was possible to study all members of the extended family. Prohibition of alcohol and drug abuse and the practice of non-violence . . . enhanced the likelihood that episodes of mania and depression could be ascertained and

diagnosed accurately.

A large collaborative study on bipolar disorder in the Amish was designed and begun in 1976. Initial studies, using five pedigrees and testing for color blindness (CB), and seventeen other markers, found no statistical evidence for linkage to Xg, color blindness, or HLA. In 1983, the group began using RFLP technology in their search for linkage and within a year Gerhard et al. reported linkage data from the same pedigree, which showed suggestive evidence of linkage of bipolar disorder to the insulin locus (INS) on the short arm of chromosome 11 (lod of 1.75 at a recombination fraction of 0.0). Within three years, they reported negative findings with 46 autosomal markers and a positive finding for linkage of bipolar disorder to the INS and Hras1 loci on the short arm of chromosome 11. The study was extensive involving 19 affected members in an 81 person pedigree. However, two other reports excluded linkage to the c-Harvey-ras-1 locus. Detera-Wadleigh et al. reported on three families of North Americans of European origin. They were able to exclude linkage to a recombination fraction of 0.15. Hodgkinson et al. also reported the absence of linkage in three Icelandic families with INS, Hras and TH. In 1988, Gill et al. presented data from one Irish family that was segregating bipolar disorder. At the markers INS and H-ras, they found no evidence for very tight linkage (recombination fraction < .005). Kelsoe, et al. reexamined the actual family in 1989 and was unable to repeat the linkage in the family itself, thus ending the claim of chromosome 11 linkage. In 1993, Baron et al. attempted to repeat the X linked findings and found no evidence for X-linkage.

Berrettini, et al. (1994) presented evidence for a linkage of bipolar phenotype to chromosome 18 based on sib pair data. Data from all 22 studied families taken in totality did not show linkage, but several large families within the sample had lod scores above 1.0. Pauls et al. (1995) were unable to replicate this finding using three large kindreds. Stine, et al. (1995) reported on 28 nuclear families with apparent unilineal descent. Using only chromosome 18 markers (31 total) in a follow-up study of Berrettini, they found a slight increase (64%) in one marker (D18S37) using a sib-pair approach. The authors then divided their sample based on whether the transmission in the small families appeared to be of paternal origin or maternal origin. They found evidence for linkage (lod = 3.51) in the paternally derived subset. However, it is not clear at all what the justification of subdividing the sample by parent-of-origin is, and even then the authors give no plausible genetic explanation for it. However, McMahon, et al. (1995) have presented evidence to suggest that there is some evidence for a maternal effect on transmission of bipolar disorder and have speculated that genomic imprinting or mtDNA may be the reason. The highest lod score without subdividing the sample was only 1.45 which is insignificant. The generation of a higher lod score by arbitrarily subdividing the set by any criteria is highly questionable. In addition, this probe with the highest lod score is over 30cM in distance from the highest lod score of Berrettini. It seems, therefore, that this study is not a replication of Berrettini.

Conclusion

Five chromosomal areas have been reported to be linked to the bipolar phenotype. They are X-chromosome (Xg), X-chromosome (CB, G6PD, coagulation factor IX), chromosome 11 (INS-HRAS1), HLA, and chromosome 18. Evidence for the X (CB, G6PD, coagulation factor IX), Xg, HLA, and chromosome 11 (INS-HRAS1) findings are lacking, leaving only the chromosome 18 findings viable at this time, although the chromosome 18 finding is yet to be replicated. This lack of replication in the field has contributed to a growing skepticism of the usefulness and reliability of the linkage study approach in psychiatric illnesses.

There are no easy answers to these controversies, but a number of observations can be gleaned from the previous linkage studies of bipolar disorder. One problem is that relatively few groups have done the vast majority of linkage studies, which has contributed to the controversy and limited progress. Another factor may be methodologic. Some investigators have argued that a higher minimum lod score is needed in order to decrease potential false positives. An additional factor is that linkage is often reported based on a minimum of data. This causes the publication of many reports that fail to meet the minimum statistical requirements. Finally, linkage is often reported as if it is the end point. The goal of this process is beyond establishing linkage. It is to find and clone a causative gene.

After 20 years and approximately 30 studies, the status of the bipolar disorder linkage field has not changed much. Yet, it still holds great promise for answering basic questions of etiology and diagnosis. The question now becomes what will the next 20 years bring. The new molecular techniques and increased quantities of markers bode well for the future. In addition, very little of the human genome has been tested, so it is quite possible that other linkages with less ambiguity are still to be found. All efforts should be concentrated on finding unambiguous linkage. The best method to do this is to use a number of very large families that can generate statistical linkage within each pedigree. If linkage is found, then efforts to tighten the linkage and find the causative gene can be initiated. Once a gene is identified, a whole new era will begin.

OTHER ATTEMPTS TO LOCALIZE PSYCHGENES

The major emphasis of psychiatric genetic research has historically been focused on schizophrenia and bipolar disorder. However, increasing attention has been paid in recent years to the genetics of a wide range of psychiatric disorders including attempts at gene localization.

The Current Status of Localizing Genes for Alzheimer Disease

The search for psychgenes in Alzheimer disease is one of the most promising areas of current research. An early finding came in 1987 when researchers reported a tentative linkage between early onset Alzheimer disease and a marker on chromosome 21. This finding came after researchers noted that individuals with trisomy 21 tended to get a dementia syndrome in their forties that was very similar to Alzheimer disease neuropathologically. This led them to test markers on chromosome 21. It is now known that some forms of early onset dementia of the Alzheimer type are due to a specific mutations in the amyloid precursor protein (APP) (Chartier-Harlin, et al., 1991; Goate, et al., 1991; Murrell, et al., 1991; Mullan, et al., 1992). The problem is that these families are apparently rare and account for less than one percent of all Alzheimer disease.

A second area of localization was reported in 1992 by Schellenberger, et al. on chromosome 14. Duijn, et al. (1994) report linkage to chromosome 14 in one early onset family, but lack of linkage in several other families. A chromosome 1 localization was recently reported using families of German-Russian origin where Alzheimer disease was early in onset. This is the third gene identified with the Alzheimer phenotype. Each of these findings requires further confirmation, but they are lending support to the idea of significant heterogeneity in broad behavioral phenotypes.

Current Status of Attempts to Localize Genes for Autism

Linkage studies offer one solution to determine if there are subsets of autism with Mendelian transmission. However, only one formal linkage study has been reported in autism. Spence et al. studied 30 blood polymorphisms for linkage, and ruled out linkage in 19. The largest lod score 1.04 was to haptoglobin (16q22). However, when the data was divided into families with male only versus families with at least one female, a lod of 1.54 to ABO on chromosome 9 was obtained. Linkage studies using families with two or more affected members offer a unique possibility to identify a DNA marker for a subset of autism. Cytogenetic abnormalities associated with autism can provide clues for chromosomal locations to attempt linkage studies. One such clue is the recent report by Ritvo et al. of an individual with a deletion on chromosome 13 associated with retinoblastoma (location 13q14), reduced esterase D activity, and autism. Because 2-5 percent of fragile X individuals have the autism phenotype, it has been speculated that FMR-1 gene may be a major cause of autism. Hallmayer, et al. (1994) studied the CGG repeat in 35 families with multiplex autism and found no CGG amplifications. Furthermore, linkage was highly negative to the FMR-1 gene. The implication is that autism is a broad behavioral phenotype with multiple genetic causes of which FMR-1 is one, but only a very minor one.

The Current Status of Localizing Genes for Alcoholism

The association of the alcoholic phenotype with an allele of the D2 dopamine receptor gene (Blum et al., 1990) generated much excitement, but the failure of replication Gelernter et al. (1991, 1993) has led to the general conclusion that the finding was due to differences between the test and control group.

The Current Status of Localizing Genes for ADHD

Cook et al. (1995) report an association between the ADHD phenotype and the dopamine transporter gene located on 5p15.3 using a 4 allele VNTR for the DAT1 gene. An association was found between the 480bp allele and the transmission of ADHD. The authors speculate that it could signify a random chance finding, a linkage disequilibrium with an as yet unknown gene near the DAT1 gene, or simply a complex association due to unknown causes. They called for further investigation.

The Current Status of Localizing Genes for Panic Disorder

Some recent attempts have been made to localize a gene for panic disorder, but have not, as yet, produced results (Mutchler et al., 1990).

The Current Status of Localizing Genes for Stuttering

Some recent attempts have been made to localize a gene for stuttering, but have, as yet, not produced results (Mellon et al., 1993).

The Current Status of Localizing Genes for Dyslexia

Early cytogenetic based studies speculated that there was a chromosomal area associated with dyslexia on chromosome 15. In 1994, Cardon et al. presented evidence suggesting a chromosomal localization to chromosome 6.

SPECULATIONS ON THE FAILURE TO LOCALIZE GENES FOR BROAD BEHAVIORAL PHENOTYPES

It is difficult to count the number of genetic disorders that have been localized by use of polymorphic DNA markers since the mid 1980's. A fact which makes it even more disappointing that there is almost no convincing evidence at this time for localization of psychgenes for any of the broad behavioral phenotypes (with the exception, perhaps, of Alzheimer disease). This failure is not due to lack of effort, because the effort has been enormous. It is not due to lack of interest by leading researchers, because many of the best and brightest have tried to tackle the genetics of the broad behavioral phenotypes.

One problem has been the continued premature publication of promising findings without clear acknowledgement of major potential problems in the findings. The failure of replication or the outright reversal of these findings has lead to a cynicism about psychiatric genetics in general. Risch (1990) has pointed out that major inconsistencies were obvious in many of the early reports of linkage in psychiatric disorders, but they were ignored in the enthusiasm. Such reversals only heighten the preexisting skepticism that is ever present regarding the genetics of behavior. These failures only serve to promote those who would reject the genetics of behavior outright on the grounds of reductionism (Rose, 1995). The most likely reason for the failure, so far, is the very nature of the broad behavioral phenotypes. There appears to be a lack of appreciation for the high likelihood of massive heterogeneity in these phenotypes. Almost all linkage approaches combine multiple small families with the assumption that little heterogeneity exists. Almost no attention is paid to what the effect would be on the linkage study paradigm if heterogeneity was at the level of 20 - 100 causative genes. Another possibility is that these disorders are truly oligogenic in nature and the strategies used so far do not take this mode of transmission into account (Craddock, et al., 1995).

Chapter 9

Special Concepts in Psychiatric Genetics

The role of genetic factors in the transmission of mental illness has been explored in some detail. A recurrent theme arising from this study is the realization that the genetics of behavior is complex. The reason for this greater complexity of behavior is simple: the brain is by far the most complex of organs. Due to the unique aspects of the brain, there are several special concepts in psychiatric genetics that must be explored in more detail.

VARIATION IN BEHAVIORAL DISORDERS AND THE SPECIAL ROLE OF DEVELOPMENTAL PLASTICITY

Diversity is a way of coping with the possible.

François Jacob

All of the important questions in psychiatric genetics can ultimately be expressed in terms of variation. How does the initial genetic endowment manifest itself throughout development and at the endpoint – the phenotype? What factors modify the genetic endowment? Can the variation be predicted? Is the variation so great in behavior that the relationship between genotype and phenotype is lost? The answers to these questions require a deeper look at the concept of variation.

Variation

Genetics is the study of variation. As such, it attempts to account for all variation in the phenotype. The total variation of a population is known as the V_P or "phenotypic variation." One of the truisms of genetics for many years has been the notion that V_P had to be due to the summation of two factors, the genetic variance V_G and the environmental variance V_E. This idea has its origins in the "nature - nurture" question that has great antiquity. This relationship is often represented by the following equation:

$$V_P = V_G + V_E$$

In other words, all phenotypic variation is due to the variation from genetic factors and the variation due to the environment. By definition and logic the genetic contribution is the difference in phenotype due to differences in the DNA between individuals. For instance, in highly inbreed lines where there are no theoretical differences in the DNA between individuals, it can be argued that no amount of the V_P can be due to genetic causes so that only the V_E can account for variation. The same argument is made for the phenotypic variance in identical human twins.

The variation due to environmental factors is much harder to isolate, since, by definition, all factors which might effect phenotypic outcome that are not genetic are environmental. These include maternal factors, nutrition, climate, or any other outside factor that may interact with the phenotypic expression. This is where the problem might have ended if it were not for the presence of a third form of variation. It has gone by a number of names including "developmental noise," "intangible variance," or “random variance.” There is a growing recognition that factors in development that are not traditionally environmental factors have a great influence on final phenotypic outcome. Many of these events involve random, stochastic processes that are amplified by chaos. This source of variation deserves to be separated from the more typical environmental sources of variation. Therefore, we can speak of variation being due to genetic, environmental, and random factors.

$$V_P = V_G + V_E + V_{DP}$$

Within this equation is the potential answer to phenotypic variation in human abnormal behavior. By analyzing the role of each variable it becomes possible to predict the relative influence of each.

Sources of Phenotypic Variation in Behavioral Disorders

Genetic endowment - The first variable that influences variation is the initial genetic endowment. However, we have seen that given a very specific genetic defect, there is often significant variation in the resulting phenotype. The cases where this is not the result seem to be examples of severe disorders that only lead to the most complete failure of behavior. When the disorders are milder, variation is the rule. This variation is so wide, that we are often left only with the recognition that a specific genetic defect X has a probability of severe disease, a probability of mild disease, and even a small possibility of no disease.

According to a pure genetic causation model, none of this variation can be due to the initial genetic endowment since the same genetic defect exists in all cases. However, even with the exact same genetic defect, there is room for variation due to genetic differences at locations other than the genetic defect in question.

Developmental Plasticity - The development of the human organism from a single cell to a viable being is a process that is understood in only the vaguest of outlines. Table One, for example, lists only the hallmark events in CNS development.

TABLE ONE:
HALLMARKS OF CNS DEVELOPMENT

Age	Event
19 days	Enlargement of anterior neural plate
24 days	Partial fusion of the neural folds
30 days	Closure of the neural tube; formation of rhombencephalon, mesencephalon, and prosencephalon

TABLE ONE: (CONT)

34 days	Formation of cerebellar plate; formation of cervical and mesencephalic flexures
38 days	Neural hypophysis forms
44 days	Olfactory evagination forms; formation of cerebral hemispheres
52 days	Optic nerve completes connection to brain
56 days	Cell migration in primitive cerebral cortex; formation of dura and pia mater
106 days	Myelination begins
168 days	Cerebral cortex exhibits typical histologic layers
196 days	Formation of prominent cerebral fissures and convolutions
280 days	Continued myelination

Through a series of cell divisions the single cell fertilized egg develops into an approximately 60,000,000,000,000 cell organism. If every division was followed by another division on a regular schedule it would take a minimum of 43 divisions to exponentially go from a single cell to a 6×10^{13} biomass. Of course this assumes every cell divides, no cells die, and every cell divides at the same rate. It is probably safe to assume that the genetic information of the 50,000 to 100,000 genes lacks the specificity to determine every cell division, every cell migration, every cell death, and other event in development. For example, when the optic nerve is studied in genetically identical individuals, plasticity is apparent (See Figure One).

The genetic endowment has a guiding and controlling effect on developmental events, but there must be wide latitude for random events in development, that are governed only by the laws of probability. Furthermore, it is assumed that the culmination of these events, in terms of phenotype, must be a chaotic process, in that the cumulative effect of these random events may have profound and unpredictable effects on the phenotype. This variation due to developmental plasticity appears to play a major role in behavior, that is not seen as clearly in other areas of genetics. One of the best examples of this type of variation is monozygotic twins with behavioral disorders. The significant variation in these twins, including even the absence of disease, is as likely due to developmental plasticity as it is to any environmental variation (Mellon, 1990).

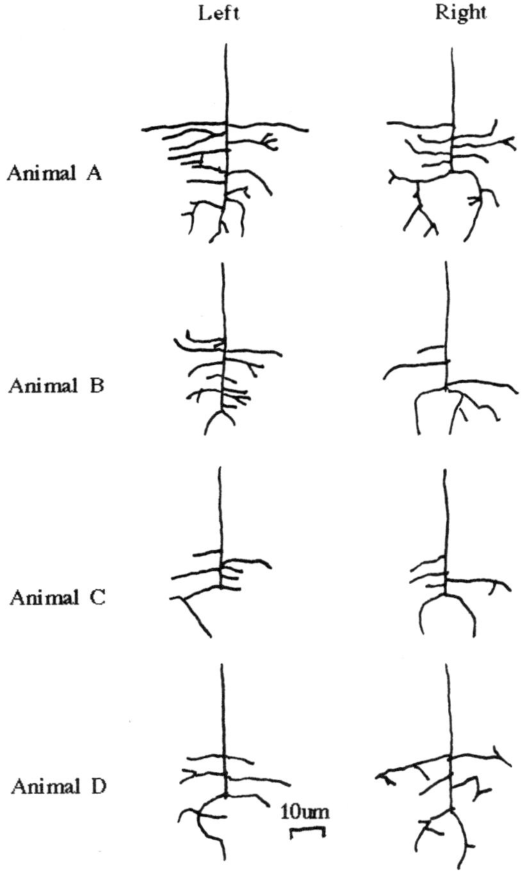

Figure One: Dendritic branching patterns of the optic ganglion neuropil in four isogenic members of *Daphnia* magna. Not only are the dendrites not symmetrical, they also vary considerably from one isogenic individual to the other. The implication is that dendritic branching is beyond specific genetic control and exhibits randomness. (After Purves and Lichtman, 1985, p. 52.)

A clinical example of this same phenomena is the case of the Genain quadruplets. These four monozygotic sisters were studied extensively by David Rosenthal and others at NIMH over a multiple year period. All four of the sisters are concordant for schizophrenia. No plausible environmental explanation has been produced to explain either their concordance or the observed clinical variation. Significant variation in a number of parameters exists even in these "identical" sisters (See Table Two).

TABLE TWO:
THE GENAIN QUADRUPLETS

Parameter	Nora	Iris	Myra	Hester
Weight at Birth	4lb 8oz	3lb 5oz	4lb 4oz	3lb 0oz
Length	17"	16"	16.5"	13.75"
Age of Onset	21	22	25	24
Years Hospitalized	7.5+	15+	1	15+
IQ	87/89	99/97	89/82	77/80
Diagnosis	Chronic Undiff Schizo	Chronic Undiff Schizo Paranoid	Schizo-affective	Chronic Undiff Schizo

Somatic mutations are another source of variation during development. Each cell duplication has an inherent risk for further genetic mutation. These somatic mutations are not passed on in the germ cells (except in the very early cells destined to become the germ cells). The importance of these somatic mutations is twofold. First, if early in development, they can lead to illness. Such early mutations lead to mosaic situations where certain percentages of cells are abnormal and some are normal. Obviously, if 100,000 cells are abnormal it has little effect, but if 30,000,000,000 cells are affected then disease may result. Second, even minor mutations in development may have some mild effect, and the results of thousands of such mutations may lead to chaotic influences that alter the ultimate phenotype.

Another example of this phenomenon is so-called developmental mutations. Evidence is growing that certain events in very early development may lead to gross alterations in organ formation often with resulting disease. These "events" do not follow genetic patterns, and in some cases appear to be random. What is suggested is that there are certain molecular steps in development that may carry certain odds of failure. In other words, these steps are completed correctly 999 out of a thousand times, but are prone to statistically fail 1 in a 1,000. The important point is that these events appear random, yet have significant phenotypic impact.

Environmental Variation - One of the major misconceptions about phenotypic variation in behavioral disorders is the claim that lack of 100 percent MZ twin concordance implies that "environmental factors" are involved in the etiology. As we have seen, developmental plasticity could account for some or all of the variation. Even if environment did play a role, it would be far more likely that it was in the form of toxins, neonatal trauma, or anoxia than in life events. It is an amazing simplicity that pervades behavioral science that assumes life events play a major role in the phenotypic variation of broad behavioral phenotypes. What is even more amazing is the virtual failure of behavioral scientists to find any of these clear cut environmental factors after a hundred years of looking. Despite this failure, the obvious examples of clear environmental factors, like alcohol damage to CNS tissue and infectious agents, are often overlooked. The reality is that it is still not clearly known what role, if any, early childhood or adulthood psychological events play in the development of abnormal human behavior, but the promotion of these events as causative in the absence of indisputable evidence only undermines the scientific basis of abnormal human behavior research.

The Effect of Variation on Final Phenotypic Outcome

Variation is present in populations from which individuals

mate. Variation is magnified during every step of development. Variation continues after birth through a combination of innate and external events. The result is that final phenotypic outcome displays a wide variation that can lead to disease without strong genetic predisposition or normality even with strong genetic loading (See Figure Two and Three).

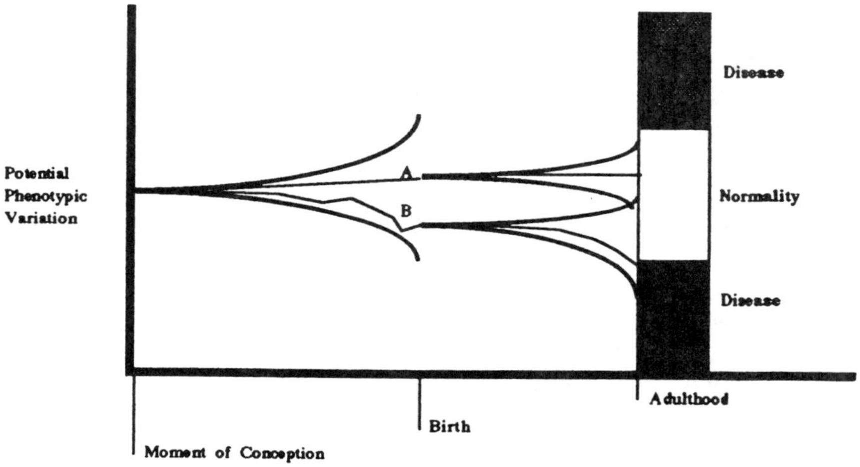

Figure Two. Phenotypic Variation with Normal Genotype. Starting at the left of the chart at the moment of conception, variation due to randomness begins. By birth, potential variation is represented by the outside lines. The course of two individuals (A) and (B) are charted to this point. It important to realize that these individuals are considered genetically identical. At birth variation continues for each individual. (A) continues along a likely path and ends up with a normal phenotype. (B) has random events in development pushing him more towards disease and with additional factors after birth ends up with a disease state.

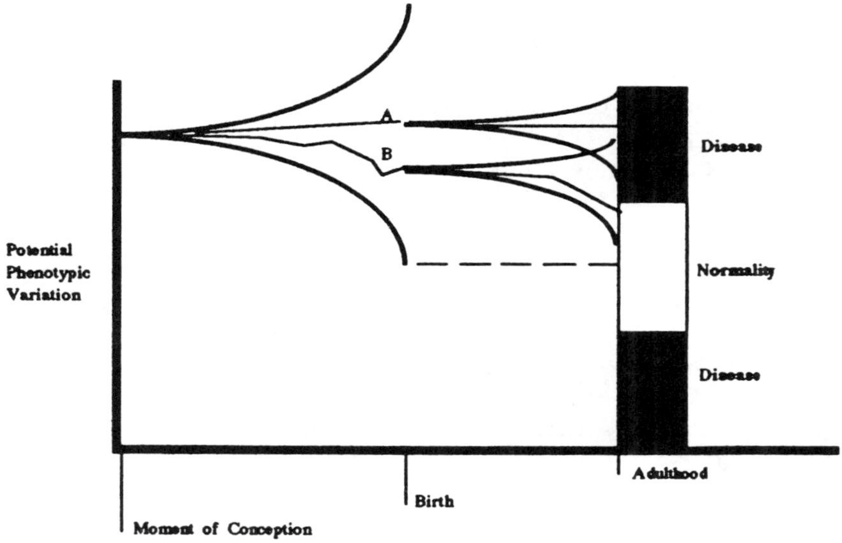

Figure Three. Phenotypic Variation with Abnormal Genotype. Starting at the left of the chart at the moment of conception., variation due to randomness begins. By birth, potential variation is represented by the outside lines. The course of two individuals (A) and (B) are charted to this point. It important to realize that these individuals are considered genetically identical. However, compared to Figure Two, these individuals have a different genetic background and have a much higher likelihood of disease. At birth variation continues for each individual. (A) continues along a likely path and ends up with an abnormal phenotype. (B) has random events in development pushing him more towards normality and with additional factors after birth ends up without a disease state.

The Evolutionary Importance of Developmental Plasticity

The factors described above result in variation in phenotypic outcome for behavioral disorders, thus reducing the relationship between genotype and phenotype. Why is this the case? What evolutionary role does it play? What forces have shaped this role? These questions deserve a proper scientific answer and the first indications of the answer are at hand.

Man, like all living organisms, can be seen as a mechanism to increase the survival of DNA. Each individual from a biologic sense serves only to perpetuate and occasionally to magnify the survival of its composite genes. Reproduction carries only the illusion of immortality for the individual, but provides the closest thing to immortality for the genes that determine the individuals fate. Under this reasoning, all behavior must have some purpose that increases the survival of genes. Altruism, for instance, is understood as behavior that may decrease the reproductive fitness of an individuals genes, but is compensated for by the survival of the common genes in a close relative. In other words, it makes sense in evolutionary terms to risk your life to save three of your siblings.

What then is the evolutionary purpose of developmental plasticity? If the phenotypic effect of a gene is not tightly linked to its genotype, then how does this contribute to the survival of genes? For instance, a gene in one individual allows for stable behavior that leads to greater reproduction. Yet, in another person (even an identical twin), the same gene can lead to aberrant behavior that reduces reproductive fitness. *The answer is that loosening the fit of genotype to phenotype creates a wider range of behavioral options, which then provides a higher likelihood that at least some form of behavior will survive.* This allows a gene to survive in the widest range of environments at only a minor cost. Developmental plasticity can thus be seen as one more layer of phenotypic variation in a system that is replete with mechanisms to extend diversity. As has been noted, evolution creates diversity, sex serves to drive phenotypic diversity, mutation increases diversity, and so does development plasticity. The power of this diversity lies in its ability to increase the survivability of a given genotype by expanding the range within which that genotype can survive. One genotype leads to numerous phenotypes thus greatly increasing the chance that many will survive and reproduce that genotype.

THE CONCEPT OF PHENOCOPIES AND ITS ROLE IN BEHAVIOR

The major premise of this work is the idea that the vast majority of abnormal human behavior is due to genetic causes. This premise contrasts with the popular view that most forms of mental illness are caused by complex environmental factors where genetics plays, at best, a minor contributing role. Ample evidence has been presented to document the central role of genetic factors. This leaves the question of what role does environment play? The first role for environment is as a playing field on which the genetic influences play out. It is in this arena that environment exerts its major influence. However, the influence is in no way causative, but instead determines the degree to which the genetic influence will be exerted. This idea contrasts greatly with the genetic-environment threshold models where genetic and environmental effects are summed and when they exceed a threshold the disorder becomes apparent. In the genetic model the environment mostly plays a protective role. In other words, environmental manipulation is a mechanism to minimize the full genetic effect of the illness. A simple example would be a person who genetically stutters who modifies his environment by speaking very slowly and avoiding stressful speaking situations gains a mastery to the point where he does not appear to be a stutterer. However, if placed in the right situation, under stress, his dysfluency may emerge again. In neither case is the environment causative of the condition, but in both cases the environment modified the condition.

Despite the genetic orientation of this work, it is readily acknowledged that there are well documented, if limited, examples of disorders of environmental origin that mimic the phenotypes of genetic disorders of behavior. Such environmental disorders are termed phenocopies. Examples include behavioral abnormalities due to exogenous teratogens, early postnatal environmental events, and later environmental events. The following are some well documented environmental effects on behavior.

Mercury

In the 1950's, the Japanese people were shocked by the outbreak of an epidemic of neurological and psychologic problems in the offspring of women in the Minimata Bay region of Japan. After a thorough epidemiologic investigation it was determined that the cause was due to methylmercury contamination in fish from a plastics plant. The behavioral consequences almost invariably involved cognitive impairment (usually mental retardation). Mercury is a classic example of a behavioral teratogen.

Phenylketonuria

When a fetus is carried by a mother afflicted with PKU, and the mother has not been treated by dietary restriction, mental retardation is the unfortunate result for the offspring. Excess phenylketones cross the placental membrane and disrupt normal development of the brain leading to mental retardation. This happens regardless of whether the offspring have PKU.

Alcohol

Growing evidence confirms that prolonged exposure of the developing fetus to ethyl alcohol results in a syndrome of physical and mental abnormalities labeled "fetal alcohol syndrome (FAS)." This syndrome is typified by facial abnormalities, intrauterine growth retardation, and cognitive impairment (with much greater risk of mental retardation). Symptoms of severe cases have been mimicked in animal models, while controversy continues about the possibility of behavioral effects in FAS in the absence of physical symptoms.

Lead

Studies have implicated serum lead levels with lowered IQ in certain high risk groups, which has resulted in great concern over the exposure of children to lead in paints and other sources. Lead appears to be a potent destructive agent of the developing CNS. What is not clear, however, is whether the behavior of ingesting lead is unrelated to initial IQ.

The Future of Behavioral Teratogenesis Research

Partly as a result of the thalidomide disaster of the 1950's, the U.S. Food and Drug Administration initiated a protocol for screening new drugs for teratogenic potential prior to approval. The current system involves a three segment process that first uses animal (usually rat) models. In the first segment the effect of the drug on fertility and overall reproductive fitness is tested. In the second segment the drugs possible teratogenic effects are studied by exposure of rats to the agent with the examination of fetuses the day prior to delivery for a range of physical structural defects. In the third segment the drug is examined for its effect on labor, delivery, and lactation.

In the 1970's Japan and the United Kingdom began to design some protocols for testing for behavioral teratogenesis. In the United States the Collaborative Behavioral Teratology Study of the 1980's was the starting point for coming to an agreement on the usefulness and appropriateness of a test for behavioral teratogenesis. So far, agreement has not occurred due to the complexity of what tests to use and how to interpret them. It appears, however, that teratogenic causation of abnormal human behavior is rare and accounts for less than three percent of all abnormal behavior.

Is Congenital Deformation a Source of Abnormal Human Behavior?

Deformation is a term that is used to describe the effect of unusual forces on a normal developing fetus. It is estimated that two percent of all fetuses have some deformation effect. Table Three lists the major deformations and their hypothesized causes. Uterine constraint can lead to unusual forces upon the developing cranium. The cranium is a common recipient of unusual forces whenever there is a problem due to the large size of the human head in relation to the body. It is generally accepted that normal brain development in embryonic development "drives" cranial size and shape, so, conversely, unusual forces on the cranium can effect the size and shape of the developing brain. However, only in the most rare and extreme examples of craniosynostosis is the underlying brain damaged to the point that it results in behavioral effects. The brain is quite malleable and functions normally even if greatly deformed in utero.

TABLE THREE:
MAJOR DEFORMATION AND PROPOSED CAUSES

Result	Proposed Cause
Joint Contracture	Immobilization & constraint
Dorsiflexed foot, clubfoot	Uterine constraint
Palsies	Nerve compression
Craniosynostosis	Limitation of suture growth by compression
Oligohydramnios Sequence	Insufficient amniotic fluid

The Role of Psychological Life Events in Creating Abnormal Behavioral Phenotypes

From ancient times people have believed that even the most serious mental illnesses could be caused by traumatic life events or even trivial life events. This misconception is unfortunately still common in modern day theories of mental illness. The amazing thing about it is how little, if any, scientific evidence there is to support this widespread belief system. Psychoanalytic and behavioral theorists have often come into direct conflict with biological and genetic views of mental illness because of this fixation on the role of environment in theories of mental illness. Only the most poorly informed individuals still believe that schizophrenia, manic-depression, and other major psychiatric disorders are due to life events. However, several other disorders retain a strong remnant of environmental bias.

Multiple Personality Disorder (MPD) - No better example exists of the quasi-intellectual result of environmental based theories than multiple personality disorder. Historically, there have been a few cases of individuals (almost exclusively women) who have presented as if they had two or more distinct personalities shifting back and forth over time. This curious phenomenon is labeled dual consciousness and is regarded as a disorder too rare to concern all but a few academicians. Since the 1970's, however, there has been an explosion in the diagnosis of MPD that is quite amazing. Studies of these individuals indicate that a high percentage report childhood abuse. It has, therefore, been concluded that childhood trauma causes this disorder. This is despite the fact that most people who have been abused do not have the disorder and only not all of the people with the disorder have been abused.

On closer inspection, virtually every case of MPD is an elaboration upon a common and better understood condition called borderline personality disorder. These individuals have mood swings, impulsiveness, suicidal tendencies, self cutting, and severe identity issues. They often over endorse any type of pathology including past pathology. Borderline personality is closely linked with hysteria and antisocial personality and is often familial. The retrospective studies used in this group prove nothing about the relationship of life events to mental illness. Only prospective studies of abused children would help answer this question, and even that type of difficult study could not determine if factors of the illness itself might predispose the child to being a victim (as for example children with mental retardation are more often victimized).

Post Traumatic Stress Disorder (PTSD) - PTSD is a poorly defined disorder that clearly overlaps with the normal human response to noxious events. This condition is typified by flashbacks to the traumatic event and nightmares of the event. It often leads to secondary gain such as disability payments. We have the capacity to learn from near fatal experiences, such as car wrecks. It is normal to "relive the event" and to ponder "what if" questions. Aversion to similar events is also normal. PTSD is the extreme of this reaction and is often due to other factors than the event itself. The condition is considered pathological when it continues for prolonged periods and interferes with normal functioning. Unfortunately, no distinction is made in this disorder about preexisting personality and mental tendencies. It appears that in many cases trauma is in the "eye of the beholder." In other words, when multiple people are traumatized at the same time as in an airline disaster or wartime, there is tremendous variation in the psychological reaction with many showing no signs of pathology at all. What this suggests is that the trauma is only a trigger for preexisting pathology. A good example is that of combat neurosis. When psychological casualties are studied during wartime, the best predictor of lack of recovery is not severity of trauma, but a positive family history of mental illness.

Brief Reactive Psychosis - This condition supposes that an otherwise normal individual is subjected to severe psychological trauma and then exhibits psychotic symptoms. As described, this condition does not exist. In reality, the individuals with this disorder have significant psychiatric problems that are "unmasked" by the trauma and not caused by it. An example is the death of a spouse with a subsequent psychotic episode in the surviving spouse. What is found on deeper inspection is that the deceased spouse compensated for and protected the spouse that survived and masked the level of pathology from others. In the absence of the spouse, the compensated behavior is revealed and appears to be new.

Chapter 10

Implications for the Practice of Psychiatry

IMPLICATIONS FOR IMPROVED DIAGNOSIS AND TREATMENT

Revolution is an overused word, but it certainly applies to the exploding knowledge base of genetics and molecular biology and its impact on the diagnosis and treatment of mental disorders. Diagnostic categories based on old dynamic theories and other outdated concepts will be replaced with a logical etiologic based diagnostic nosology derived from specific molecular mechanisms. Psychiatric disorders are increasingly finding their places in the "morbid anatomy" of the human genome. As is usually the case, treatment improvements are slower in coming, but hold the promise of being specific, etiologic based interventions with the greatest possible impact and benefit. Of the several reasons for studying the inheritance of abnormal behavior, none is more important than the possibility for potential improvement in diagnosis and treatment.

As a field, psychiatric genetics has certain goals for the future. The first task is to determine what, if any, component of a behavioral disorder is genetic and to fully explore the implications of the findings. When a disorder is genetic, the goal is to determine the exact mode of genetic transmission. The next step is to use the genetic information about a disorder as a tool to find the causation of the disorder at the molecular level. Finally, the goal is to use the genetic etiology knowledge to find better treatments.

Data collection is of central importance in the diagnosis of genetically determined behavior abnormalities. Traditional methods of data collection must be expanded to include detailed family histories, genetic lab tests, and dysmorphology based physical examinations. Together these items will allow for a complete appraisal of the individuals condition within our current knowledge base. At a minimum, they will provide a historical document that may help the individual or his offspring in the future.

A Genetic Oriented Family History

A family history is of critical importance and should never be overlooked in any assessment of abnormal behavior. It is important that a true genetic family history be taken, not the sort of family relationship charts that are often substituted in social work histories, but a complete family genetic history. There are accepted standards of making family pedigrees that should be followed (Bennett, et al., 1995). It is important to realize that probands and other family members are not good judges of whether other family members are affected. It is usually not possible to interview all family members, but at least there should be no delusions about the validity of family reports. In general, however, errors are usually made in leaving out affected individuals than in false positives. Other important findings that should always be addressed include consanguinity, miscarriages, and suicides.

The Physical Exam: The Psychiatrist as Dysmorphologist

It is no surprise that errors in embryogenesis are often reflected in recognizable patterns of maldevelopment. The study of such errors in the individual is known as dysmorphology. The premise of dysmorphology is that the developing individual is subject to a number of problems in growth or morphogenesis. One source of problem is malformation due to the poor formation of tissue from some earlier event (mutation) that leads to a sequence of events resulting in one or more physically apparent abnormalities. Another problem is deformation, which results from unusual forces impinging upon the normal developing fetus. Finally, disruption can occur when normal tissue is destroyed. Most physical exams completely ignore all but the most blatant examples of abnormalities. Each of these clues can play an essential role in the diagnosis of behavioral disorders, especially mental retardation. It is beyond the scope of this work to try to describe the adequate dysmorphology exam. In the references is a list of several of the most useful guides to dysmorphology examination.

Using Genetic Related Lab Tests

Several specific lab tests are available to aid in the diagnosis of behavioral disorders. The karyotype is an analysis of chromosomal structure, which can detect aneuploidy, deletions, and other chromosomal abnormalities. The metabolic screen is a test based on urine and/or blood, which can detect a wide range of inborn errors of metabolism. Other specific tests include direct DNA screening for certain disorders, ceruloplasmin levels, and long chain fatty acids tests.

Treatment: General Principles

The growing recognition of the central role of genetics in the development of human behavioral disorders is leading to a new understanding about the efficacy of traditional treatment modalities. A new synthesis is being formed that takes the best of the old treatments and modifies them with the best of the new. What is emerging is a holistic view of treatment that accounts for every aspect of genetics, development, and environmental insult. In this view there is a constructive, if slightly changed, role for all the traditional treatment modalities. The knowledge gained from the study of the genetic basis of abnormal human behavior is of little value unless it translates into improvements in the treatment of individuals with mental illness.

Improvement in treatment of mental illness can be divided into at least three levels; prevention, symptomatic management, and cure. Strategies based on prevention in the past have been mostly limited to eugenic notions, which we have seen have not only exaggerated their usefulness, but fail to help the individual. In more recent times, prenatal diagnosis with selective abortion has become the major form of prevention of known genetic mental illnesses. This strategy also has many drawbacks and is highly controversial at best. True prevention, such as gene replacement, is only in its infancy, but some strategies based on polar body analysis have shown some promise. Much more success has been had in the area of symptomatic management of human behavioral abnormalities. Through the use of psychopharmacology, many symptomatic improvements have been made. Learning strategies and psychotherapy have a less important role. Perhaps the most unsatisfactory aspect of treatment is the very few instances in which a cure can be claimed.

The antagonism between biological psychiatry and psychotherapy is a falsehood. No one who has dealt with the mentally ill in any serious capacity can deny the importance of the therapeutic relationship. What is contested, however, are psychotherapeutic tautologies usually based on some etiologic theory. In these well meaning but erroneous theories there is usually a putative environmental event in the form of some life event or series of life events that lead to the psychopathology. Therapy is based on archeological recreation of these events and a rebuilding of an alternative construct. This conflicts directly with the assertion that etiology is primarily genetic. The role of psychotherapy in the genetic model is limited to therapeutic alliance, education, and, as Guze has pointed out, rehabilitation.

Clearly, the genetic theory of abnormal human behavior favors the physical therapies. Pharmacotherapy is the cornerstone of treatment in the genetic model at this point. The use of molecular treatments fits well with the idea that molecular defects are at the core of most of these disorders. However, most drugs used at this time are less than satisfactory because of side effects and nonspecificity of effects.

Treatment: The Psychiatrist as Genetic Counselor

With the advance in understanding of the genetic aspects of psychiatric disorders comes an increasing need for genetic counseling for psychiatric disorders. Often family members want to know if they or their children will be at risk. They will look to the psychiatrist for answers. All genetic counseling is fraught with ethical issues and complex risk assessments, it is not a task to be taken lightly. However, psychiatrists should have the skills and knowledge to provide specific behavioral genetic counseling. With the growing evidence for the genetic basis of mental illness it is essential that psychiatrists begin to take responsibility for the genetic counseling of their patients. It is irresponsible to cast this duty off to other specialties. With a little training and understanding the psychiatrist is the ideal person to deal with genetic counseling as it relates to behavior. Psychiatrists know more about the behavioral disorders than non-psychiatrists. Psychiatrists certainly are in a unique position to deal with the counseling aspects.

Clients asking for genetic counseling can be the patients themselves, family members (affected and non-affected) and potential spouses. Society at large, while effected by these issues, has very little role in the nature of genetic counseling. Counseling should always be directed to needs and desires of the individuals seeking counseling, and societal needs should not enter into the equation. The goals of genetic counseling involve providing the best and most accurate information to the client so that the client may then make informed decisions. The goal is not to persuade or dissuade anybody from any particular decision. A person competent to provide genetic counseling has no problem understanding how one family would choose to have a child with trisomy 21 while another would choose abortion. Each family will do what is right for them. There are few absolutes. Alleviating anxiety by providing information is the main goal of genetic counseling. All is predicated on the belief that difficult decisions are made easier, not harder, by the addition of knowledge. The main form of knowledge provided are risk estimates. These can be very precise as in Mendelian traits, or less specific risk assessments based on family studies - the so-called "empiric risk." At the present time, most risk assessments in behavioral disorders are empiric.

Technique of Psychiatric Genetic Counseling

The first requirement of proper genetic counseling is an accurate diagnosis and family history. Psychiatrists are the obvious choice for making behavioral diagnoses and should be able to construct detailed family histories focused on behavioral problems. The next task is to assess the level of understanding of the clients so that information can be presented on a level that is understandable to them. Determination of risk calculations is the next order of business. It is very important to emphasize the empirical basis of most risk estimates. Also, it is important to emphasize the legitimate areas of uncertainty where they exist. Studies have shown that the vast majority of clients want to know the most important details as soon as possible (Mikkelsen et al., 1981). They do not necessarily want to know or need to know everything at the first meeting. Some studies suggest that they will often forget most of what is told to them, even things as basic as the diagnosis and the recurrence risk. At the initial meeting there may be some bitter disappointment. It is important to allow an appropriate expression of feelings. It is absolutely unacceptable to rush this meeting or report the findings with no awareness of the implications for the family and patient. After this meeting, is it essential to provide written documentation of risks and other important findings and to send a copy of this document to the clients. Failure to do this will allow for may misconceptions since studies have shown that clients do not absorb most of what is said in genetic counseling. Finally, leave the lines of communication open. A follow-up meeting is often very helpful, if not essential.

Problems in the Counseling of Behavioral Genetic Disorders

In all forms of counseling, there is the issue of transference, which must be dealt with. The possibility also exists that the counselor is influencing the family with subtle gestures (smiles, head nods, etc.) to a decision that he or she thinks is best, but is not an independent decision of the family. It is critical to remember that many families are overwhelmed by the information given them and will ask, "what would you do?" In this and other ways, families will often try to have the "expert" make the decision and relieve them of the duty. Several studies have shown that much of what is discussed in genetic counseling is often forgotten by the family. Reiss and Menasche found that none of 39 families counseled about their child's recurrence risk for a heart defect were aware of the recurrence risk. Only 9 of the 39 remembered it when counseled again and questioned later! Leonard et al. studied the communication between counselor and families using 61 families. They found that in about half the families there was a significant breakdown in communication including five families that felt they had never been counseled, six that felt the counseling was in error, and 14 more families that either could not remember or comprehend the details of the counseling. Others have pointed out that even the basic words used in counseling may have completely different meaning to the patient (Fresco & Silvestre, 1981). Many clinicians have partially solved this problem by giving each family a detailed letter containing the specific findings and recurrence risks.

From early in the history of genetic counseling it became apparent that many parents reacted to genetic risks by denying the illness and the potential risk. Families often quickly dropped out of the clinic after receiving difficult results. Others, when contacted later, implied that they were never told there was any risk. Some simply stated that the results are wrong and they or their child was doing well. Studies have shown that the presence of a genetic defect in a family can have tremendous influence on individuals self worth and self view as well as in the disruption of the family structure. Psychiatrists, more than anyone else, should be able to see denial forming and help the families process through the grieving and come to some terms with the reality of the situation. Another reason that psychiatrists should be involved in genetic counseling of behavior disorders is that they are in a keen position to see when grief reactions start or an adjustment reaction sets in and to provide effective treatment.

Appendix A

Behavioral Genetics Related Service Organizations

Alzheimer's Disease and Related Disorders Association, Inc.
900 N. Michigan Avenue, Suite 1000
Chicago, IL 60611
(312) 335-8700

Autism Society of America
8601 Georgia Ave. # 503
Silver Spring, MD 20910
(301) 565- 0433

Depression and Related Affective Disorders Association, Inc.
Johns Hopkins Hospital
Meyer 4-181
601 North Wolfe Street
Baltimore, MD 21205
(301) 955-4647

Huntington Disease Society of America, Inc
140 West 22nd St
New York, NY 10011-2420
(212) 242-1968

National Fragile X Foundation
1441 York St, Suite 215
Denver, CO 80206

Neurofibromatosis, Inc
8855 Annapolis Road, Suite 110
Lanham, MD 29706
(800) 442-6825

Orton Dyslexia Society
Cluster Bldg, suite 382
8600 La Salle Rd
Baltimore, MD 21204
(410) 296-0232

Tuberous Sclerosis Association of America
P.O. Box 1305
Middleboro, MA 02370
(617) 947-8893

Tourette Syndrome Association, Inc.
42-40 Bell Boulevard
Bayside, NY 11361
(718) 224-2999

Turner Syndrome Society of the US
15500 Wayzata Blvd. #768-214
Wayzata, MN 55391
(612) 475- 9944

Williams Syndrome Association
P.O. Box 5297
Ballwin, Missouri 63022
(314) 227-4411

Wilson Disease Association
P.O. Box 75324
Washington, D.C. 20013
(703) 636-3003

References

EPIGRAPH

Morgan, Thomas Hunt (1913) *Heredity and Sex*. New York. Columbia University Press, p. vi.

A SELECTED LISTING OF OVERVIEWS OF PSYCHIATRIC GENETICS

Myerson, A. (1925) *Inheritance of Mental Diseases*. Baltimore. Williams and Wilkins.

Slater, E. (1938) "Mental Disease, Heredity". In *British Encyclopedia of Medical Practice.* Edited by H. Rolleston, Vol 8, pp. 552-63. London. Butterworth.

Slater, E. (1953) "Psychiatry." In *Clinical Genetics*, Edited by A. Sorsby. London. Butterworth & Co.

Kallmann, F.J. (1953) *Heredity in Health and Mental Disorders: Principles of Psychiatric Genetics in Light of Comparative Twin Studies.* New York. W.W. Norton.

Kallmann, F.J. (Editor) (1962) *Expanding Goals of Genetics in Psychiatry*. New York. Grune and Stratton.

Rosenthal, D. (1970) *Genetic Theory and Abnormal Behavior.* New York. Mcgraw-Hill.

Slater, E., Cowie, V. (1971) *The Genetics of Mental Disorders*. London. Oxford University Press.

Fieve, R.R., Rosenthal, D., Brill, H. (Editors) (1975) *Genetic Research in Psychiatry*. Baltimore. Johns Hopkins University Press.

Worden, F.G., Childs, B., Matthysse, S., Gershon, E.S. (1976) "Frontiers of Psychiatric Genetics." *Neurosciences Research Program Bulletin* 14:6-87.

Sperber, M.A., Jarvik, L.F. (1976) *Psychiatry and Genetics*. New York. Basic Books.

Tsuang, M.T., Vandermey, R. (1980) *Genes and the Mind: Inheritance of Mental Illness*. New York. Oxford University Press.

Gershon, E.S., Matthysse, S., Breakefield, X.O., Ciaranello, R.D. (Editors) (1981) *Genetic Research Strategies in Psychobiology and Psychiatry*.

Pacific Grove, Calif. Boxwood Press.
Henderson, N.D. (1982) "Human Behavior Genetics." *Annual Review of Psychology* 33:403.
McBroom, P. (1982) *Behavioral Genetics.* New York. Research and Education Association.
Kety, S.S., Rowland, L.P., Sidman, R.L., Matthysse, S.W. (Editors) (1983) *Genetics of Neurological and Psychiatric Disorders.* New York. Raven.
McGuffin, P. (1984) "Genetic Influences on Personality Neurosis and Psychosis." In *Scientific Principles of Psychopathology* Edited by P. McGuffin. London. Grune and Stratton.
Ranier, J. (1985) "Genetics and Psychiatry". In *Comprehensive Textbook of Psychiatry IV.* Baltimore: Williams and Wilkins, pp. 25-42.
Sakai, T., Tsuboi, T. (1985) *Genetic Aspects of Human Behavior.* Tokyo. Igaki-shoin Ltd.
Nurnberger, J.I., Goldin, L.R., Gershon, E.S. (1986) "Genetics of Psychiatric Disorders". In *Medical Basis of Psychiatry.* Edited by G. Winokur & P. Clayton. Philadelphia. W.B. Saunders Co: Philadelphia, pp. 486-521.
Gershon, E.S., Cloninger, C.R., Propping, P., DeLisi, L.E., Hippius, H. (1987) "Genetic Research in Psychiatry." *Journal of Psychiatric Research* 21: 333-639.
Reider, R.O., Kaufman, C.A. (1988) "Genetics", In *Textbook of Psychiatry.* Edited by J.A. Talbot, R.E. Hales, S.C. Yudofsky. Washington, DC. American Psychiatric Press, pp. 33-65.

SELECTED LISTING OF WORKS RELATED TO PSYCHIATRIC GENETICS

Fuller, J.L., Thompson, W.R. (1960) *Behavior Genetics.* New York. Wiley.
Vandenberg, S.G. (Editor) (1965) *Methods and Goals in Human Behavior Genetics.* New York. Academic Press.
Parsons, P.A. (1967) *The Genetic Analysis of Behaviour.* London. Methuen.
Vandenberg, S.G. (Editor) (1968)*Progress in Human Behavior Genetics.* Baltimore. Johns Hopkins Press.
Ehrman, L., Omenn, G.S., Caspari, E. (Editors) (1972) *Genetics, Environment, and Behavior.* New York. Academic Press.
Wilson, J.R. (1973) *Behavioral Genetics: Simple Systems. A Symposium held March 4-5, 1971 at the University of Colorado.* Boulder, Colorado. Colorado Associated University Press.
McClearn, G.E., DeFries, J.C. (1973) *Introduction to Behavioral Genetics.* New

York. Freeman.
Van Abeelen, J.H.F. (Editor) (1974) *The Genetics of Behaviour.* Amsterdam. North-Holland.
Schaie, K.W., Anderson, V.E., McClearn, G.E., Money, J. (1975) *Developmental Human Behavior Genetics: Nature-Nurture Redefined.* Lexington, Massachusetts. Lexington Books.
Kaplan, A.R. (Editor) (1976) *Human Behavior Genetics.* Springfield, Illinois. Charles C. Thomas.
Schmid, W., Nielson, J. (Editors) (1981) *Human Behavior and Genetics: Proceedings of the Symposium of the European Society of Human Genetics held at the University of Zurich, Switzerland, March 26-28, 1981.* Amsterdam. Elsevier.
Ionasescu, V., Zellweger, H. (1983) *Genetics in Neurology.* New York. Raven Press.
Hay, D.A. (1985) *Essentials of Behaviour Genetics.* Melbourne. Blackwell Scientific Publications.
Jones, K.L. (1988) *Smith's Recognizable Patterns of Human Malformation.* Fourth Edition. Philadelphia. W.B. Saunders Co.
Johnson, W.G. (Editor) (1989) *Neurogenetic Diseases.* Neurologic Clinics. Philadelphia. W.B. Saunders Company.
Abel, E.L. (1989) *Behavioral Teratogenesis and Behavioral Mutagenesis: A Primer in Abnormal Development.* New York. Plenum Press.

CHAPTER ONE:
THE RANGE OF ABNORMAL BEHAVIOR IN MAN

General References

American Psychiatric Association (1987) *Diagnostic and Statistical Manual of Mental Disorders*, third edition , revised. Washington, DC. American Psychiatric Association.
Pincus, J.H., Tucker, G.T. (1985) *Behavioral Neurology.* New York. Oxford University Press.
Wender, P.H., Klein ,D.F. (1981) *Mind, Mood, and Medicine: A Guide to the New Biopsychiatry.* New York. Farrar, Straus, Giroux.

CHAPTER TWO:
BASIC PRINCIPLES OF GENETICS AND THEIR IMPLICATIONS

General References

Abel, E.L. (1989) *Behavioral Teratogenesis and Behavioral Mutagenesis.* New York. Plenum.

Bateson, G. (1979) *Mind and Nature: A Necessary Unity.* New York. Dutton.

Dawkins, R. (1976) *The Selfish Gene.* New York. Oxford.

Dawkins, R. (1986) *The Blind Watchmaker.* New York. Oxford.

Gribben, J., Gribben, M. (1988) *The One Percent Advantage: The Sociobiology of Being Human.* New York. Blackwell.

Basic Genetic Concepts

Jacob, F. (1973) *The Logic of Life: A History of Heredity.* New York. Pantheon.

Stubbe, H. (1972) *History of Genetics.* Cambridge. MIT Press.

Mendel, G. (1865) "Experiments in Plant Hybridization." In *Classic Papers in Genetics.* Edited J.A. Peters (1959) New Jersey. Prentice-Hall, pp. 1-19.

Sutton, W. S. (1903) "The Chromosomes in Heredity." In *The Chromosome Theory of Inheritance.* Edited B.R. Voeller. New York (1968) Appleton-Century, pp. 157-174.

Tijo, J.H., Levan, H. (1956) *Hereditas* 42:1-6.

DNA

Lewin, B. (1987) *Genes III.* Third edition. New York. John Wiley and Sons.

The Genetic Code

Woese, C. (1967) *The Genetic Code: The Molecular Basis for Genetic Expression.* New York. Harper and Row.

Mutation

Wallace, B. (1970) *Genetic Load: Its Biological and Conceptual Aspects.* Englewood Cliffs. Prentice-Hall.

Genetic Factors in Human Illness: Mendelian Patterns of Inheritance

Hoffman, E.P., et al. (1988) Characterization of Dystrophin in Muscle-biopsy Specimens from Patients with Duchenne's or Becker Muscular Dystrophy. *NEJM* 318:1363-1368.

Genetic Factors in Human Illness: Non-Mendelian Patterns of Inheritance

Predisposition to Disease

Garrod, A.E. (1931) *Inborn Factors in Disease: An Essay*. London. Oxford University Press.

Simple versus Complex Diseases

Wallace, D.C. (1995) Mitochondrial DNA Variation in Human Evolution, Degenerative Disease, and Aging. *AJHG* 57:201-23.

How to Explain the Genetic Basis of Low Recurrence Risk Disease

Penrose, L. (1953) The Genetical Background of Common Diseases. *Acta Genet.* 4:257-65.

Edwards, J.H. (1960) The Simulation of Mendelism. *Acta Genet.* 10:63-70.

Falconer, D.S. (1965) The Inheritance of Liability to Certain Diseases, Estimated From the Incidence Amongst Relatives. *Ann. Hum. Genet.* 29:51-71.

Carter, C.O. (1965) The Inheritance of Common Congenital Malformations. *Prog. Med. Genet.* 4:59-84.

Kartagener, M. (1933) Zur Pathogenese der Bronchiektasien: Bronchiektasien bei Situs Viscerus Inversus. *Beitr. Klin. Tuberk.* 83:489-501.

Feldman, W. (1935) Transposition of the Viscera Behaving as a Mendelian Recessive with Congenital Absence of the Appendix. *Proc. Roy. Soc. Med.* 28:753-56.

James, J.W. (1971) Frequency of Relatives for an All-or-None Trait. *Ann. Hum. Genet.* 35:47-49.

Cavenee, W.K., Dryja, T.P., et al. (1983) Expression of Recessive Alleles by Chromosomal Mechanisms in Retinoblastoma. *Nature* 305:779-784.

Newman, T.B. (1985) Etiology of Ventricular Septal Defects: An Epidemiological Approach. *Pediatrics* 76:741-749.

Kurnit, D.M., Layton, W.M., Matthysse, S. (1987) Genetics, Chance, and Morphogenesis. *AJHG* 41:979-95.

Special Topics in Genetics with Relevance for Behavior

Medawar, P.B. (1957) *The Uniqueness of the Individual.* London. Methuen.
Williams, R.J. (1956) *Biochemical Individuality.* New York. John Wiley.
Sutherland, G.R., Richards, R.I. (1992) Invited Editorial: Anticipation Legitimized: Unstable DNA to the Rescue. *AJHG* 51:7-9.
Wallace, B. (1968) *Topics in Population Genetics.* New York. Norton.
Weismann, A. (1893) *The Germ Plasm: A Theory of Heredity.* London. Scott.
Pearson, K. (1903) On Inheritance in Man. *Biometrika.* 1.

CHAPTER THREE: THE BEHAVIORAL CONSEQUENCES OF CHROMOSOME ABNORMALITIES

Nomenclature

Tijo, J.H., Levan, H. (1956) *Hereditas* 42:1-6.

Behavioral Consequences of the Loss of a Chromosome

45,XO Turner Syndrome

Turner, H.H. (1938) A Syndrome of Infantilism, Congenital Webbed Neck, and Cubitus Valgus. *Endocrinology* 23:566.
Alexander, D., Money, J. (1966) Turner's Syndrome and Gerstmann's Syndrome: Neuropsychological Comparisons. *Neuropsychologia* 4:265.
Money, J., Alexander, D. (1966) Turner's Syndrome: Further Demonstration of the Presence of Specific Congenital Deficiencies. *J. Med. Genet.* 3:47-48.
Garron, D.C. (1970) Sex-linked Recessive Inheritance of Spatial and Numerical Abilities and Turner's Syndrome. *Psychol. Rev.* 77:147.
Garron, D.C. (1977) Intelligence Among Persons with Turner's Syndrome. *Beh. Genet.* 7:105-127.
Nielsen, J., Sillesen, I. (1981) "Mental aspects in Turner syndrome." In Schmid, W. & Nielsen, J. (editors) *Human Behavior and Genetics.* Amsterdam. Elsevier, pp. 133-150.
Christensen, A-L., Nielses, J. (1981) "A Neuropsychological Investigation of

17 Women with Turner's Syndrome - A.R. Luria's Theory Applied to Turner's Syndrome." In Schmid, W. & Nielsen, J. (editors) *Human Behavior and Genetics*. Amsterdam. Elsevier, pp. 151-166.

Nyborg, H., Nielsen, J. (1981) "Sex Hormone Treatment and Spatial Ability in Women with Turner's Syndrome." In Schmid, W. & Nielsen, J. (editors) *Human Behavior and Genetics*. Amsterdam. Elsevier, pp. 167-182.

McCauley, E., Sybert. V.P., Ehrhardt, A.A. (1986) Psychosocial Adjustment of Adult Women with Turner Syndrome. *Clinical Genetics* 29:284-290.

Behavioral Consequences of the Addition of a Chromosome

Jacobs, P.A., Strong, J.A. (1959) *Lancet* 1:710-712.

Edwards, J.H.,Harnden, D.G., Cameron, A.H., et al. (1960) A New Trisomic Syndrome. *Lancet* 1:787.

Smith, D.W., Patau, K. et al. (1960) A New Autosomal Trisomy Syndrome. *J. Pediatr.* 57:338.

Caspersson, T., et al. (1972) Four patients with Trisomy 8 Identified by the Fluorescent and Giemsa Banding Techniques. *J Med Genet* 9:1-7.

Trisomy 21

Lejeune, J.,Gautier, M., Turpin, R. (1959) *C. R. Acad Sci Paris.* 248:1721-1722.

Owens, D., Dawson, J.C., Lesin, S. (1971) Alzheimers Disease in Down Syndrome. *Am. J. Ment. Def.* 75:606-612.

XXY Klinefelter Syndrome

Klinefelter, H.F., Reifenstein, E.C., Albright, F. (1942) Syndrome Characterized by Gynecomastia, Aspermatogenesis, without A-Leydigism, and Increased Excretion of Follicle-stimulating Hormone. *J. Clinical Endocrinol.* 2:615.

Hambert, G. (1966) *Males with Positive Sex Chromatin. An Epidemiological Investigation Followed by Psychiatric Study of Seventy-five Cases.* Akademiförlaget, Göteborg.

Nielsen, J., Sorensen, A., Theilgaard, A., et al. (1969) A Psychiatric-psychological Study of 50 Severely Hypogonadal Male Patients, Including 34 with Klinefelter's Syndrome, 47XXY. *Acta Jutlandica*, Munksgaard, Copenhagen.

Ratcliffe, S.G., Tierney, I., Smith, L., Callan, S. (1981) "Psychological and Educational Progress in Children with Sex Chromosome Abnormalities in the Edinburgh Longitudinal Study." In Schmid, W. & Nielsen, J. (editors) *Human Behavior and Genetics*. Amsterdam. Elsevier, pp. 31-43.

Sorensen, K., Sorensen, A.M., Nielsen, J. (1981) "Social and Psychological Development of Adolescents with Klinefelter's Syndrome." In Schmid, W. & Nielsen, J. (editors) *Human Behavior and Genetics*. Amsterdam. Elsevier, pp. 45-63.

Theilgaard, A. (1981) "Childhood Experiences Reported by XYY and XXY Men." In Schmid, W. & Nielsen, J. (editors) *Human Behavior and Genetics*. Amsterdam. Elsevier, pp. 65-74.

Theilgaard, A. (1981) "The Personalities of XYY and XXY Men." In Schmid, W. & Nielsen, J. (editors) *Human Behavior and Genetics*. Amsterdam. Elsevier, pp. 75-84.

Nyborg, H., Nielsen, J. (1981) "Spatial Ability of Men with Karyotype 47,XXY, 47, XYY, or Normal Controls." In Schmid, W. & Nielsen, J. (editors) *Human Behavior and Genetics*. Amsterdam. Elsevier, pp. 85-106.

Stewart, D.A., Netley, C.T., Park, E. (1982) Summary of Clinical Findings of Children with 47,XXY, 47,XYY, and 47,XXX Karyotypes. *Birth Defects: Original Article Series* 18:1-5.

XYY Phenotype

Sandberg, A.A., Koepf, G.F., Ishihara, T., Hauschka, T.S. (1961) An XYY Human Male. *Lancet*, ii, 488-489.

Jacobs, P.A., Brittain, R.P., McClermont, W.F. (1965) Aggressive Behavior, Mental Subnormality and the XYY male. *Nature* 208: 1351.

Price, W.H., Whatmore, P.B. (1967) Behavior Disorders and the Pattern of Crime Among XYY Males Identified at a Maximum Security Hospital. *Br. Med. J.* 1:533-536.

Court Brown, W.M. (1968) Males With an XYY Sex Chromosome Complement. *J. Med Genet* 5:341.

Elkington, J.R. (1968) Sex Chromosomes and Crime. *Ann. Intern. Med.* 69:399-401.

Heller, J.H. (1969) Human Chromosome Abnormalities as Related to Physical and Mental Dysfunction. *J. Hered.* 60:239-248.

Shah, S. (1970) A Report on the XYY Chromosomal Abnormality. *Center for Studies of Crime and Delinquency. Rockville Maryland. National Institute of Mental Health. Public Service Publications*, pp. 1-55.

Jarvik, L.F., Klodin, V., Matsuyama, S.S. (1973) Human Aggression and the Extra Y Chromosome - Fact or Fantasy ? *Am. J. Psychol.* 28:674.
Christensen, A.L., Nielsen, J. (1973) Psychological Studies of Ten Patients with the XYY Syndrome. *Brit. J. Psychiatry* 123:219-221.
Hook, E.B. (1973) Behavioral Implications of the Human XYY Genotype. *Science* 179:139-150.
Witkin, H.A., Mednick, S.A., Schulsinger, F., et al. (1976) Criminality in XYY and XXY Men. The Elevated Crime Rate of XYY Males is not Related to Aggression. It may be Related to low Intelligence. *Science* 193: 547-555.
Ratcliffe, S.G., Tierney, I., Smith, L., Callan, S. (1981) "Psychological and Educational Progress in Children with Sex Chromosome Abnormalities in the Edinburgh Longitudinal Study." In Schmid, W. & Nielsen, J. (editors) *Human Behavior and Genetics.* Amsterdam. Elsevier, pp. 31-43.
Theilgaard, A. (1981) "Childhood experiences reported by XYY and XXY Men." In Schmid, W. & Nielsen, J. (editors) *Human Behavior and Genetics.* Amsterdam. Elsevier, pp. 65-74.
Theilgaard, A. (1981) "The Personalities of XYY and XXY Men." In Schmid, W. & Nielsen, J. (editors) *Human Behavior and Genetics.* Amsterdam. Elsevier, pp. 75-84.
Nyborg, H., Nielsen, J. (1981) "Spatial Ability of Men with Karyotype 47,XXY, 47, XYY, or Normal Controls." In Schmid, W. & Nielsen, J. (editors) *Human Behavior and Genetics.* Amsterdam. Elsevier, pp. 85-106.
Stewart, D.A., Netley, C.T., Park, E. (1982) Summary of Clinical Findings of Children with 47,XXY, 47,XYY, and 47,XXX Karyotypes. *Birth Defects: Original Article Series* 18:1-5.
Leonard, M.F., Sparrow, S. (1986) Prospective Study of Development of Children with Sex Chromosome Anomalies: New Haven Study. IV. Adolescence. *Birth Defects: Original Article Series* 22:221-249.

XXX Syndrome

Ratcliffe, S.G., Tierney, I., Smith, L., Callan, S. (1981) "Psychological and Educational Progress in Children with Sex Chromosome Abnormalities in the Edinburgh Longitudinal Study." In Schmid, W. & Nielsen, J. (editors) *Human Behavior and Genetics.* Amsterdam. Elsevier, pp. 31-43.
Stewart, D.A., Netley, C.T., Park, E. (1982) Summary of Clinical Findings of Children with 47,XXY, 47XYY, and 47,XXX Karyotypes. *Birth Defects: Original Article Series* 18:1-5.

2q37- Syndrome

Wilson, L.C., Leverton, K., et al. (1995) Brachydactyly and Mental Retardation: an Albright Hereditary Osteodystrophy-like Syndrome Localized to 2q37. *AJHG* 56:400-407

4p- Syndrome

Gandelman, K.Y., et al. (1992) Molecular Definition of the Smallest Region of Deletion overlap in the Wolf-Hirschhorn Syndrome. *AJHG* 51:571-578.

5p- Cri-du-chat

Gersh, M., Goodart, A., et al. (1995) Evidence for a Distinct Region Causing a Cat-Like Cry in Patients with 5p Deletions. *AJHG* 56:1404-1410.

Williams Syndrome

Williams, J.C.P., Barratt-Boyes, B.G., Lowe, J.B. (1961) Supravalvular Aortic Stenosis. *Circulation* 24:1311-18.

Lowery, M.C., Morris, C.A., et al. (1995) Strong Correlation of Elastin Deletions, Detected by FISH, with Williams Syndrome: Evaluation of 235 patients. *AJHG* 57:49-53.

11q- Jacobsen Syndrome

Penny, L.A., Dell'Aquilla, M., et al. (1995) Clinical and Molecular Characterization of Patients with Distal 11q Deletions. *AJHG* 56:676-683.

13q- Syndrome

Brown, S., Russo, J. (1995) The 13q- Syndrome: The Molecular Definition of a Critical Deletion Region in Band 13q32. *AJHG* 57:859-866.

15q11-13 - Prader-Willi Syndrome and Angelman Syndrome

Cassidy, S.B., Ledbetter, D.H. (1989) "Prader-Willi Syndrome." In Johnson, W.G. (editor): *Neurogenetic Diseases.* Neurologic Clinics. Philadelphia. W.B. Saunders Company, pp. 37-54.

Burd, L., Vesely, B., et al. (1990) Prevalence Study of Prader-Willi Syndrome in North Dakota. *AJMG* 37: 97-99.

Hall, J. (1995) The Small Nuclear Ribonucleoprotein- associated Polypeptide N (SNRPN) gene in Prader-Willi and Angelman Syndromes. *Growth, Genetics, and Hormones* 11:12

16q- Syndrome

Casamaaima, G. (1990) Deletion of 16q with Prolonged Survival and Unusual Radiographic Manifestations. *AJMG* 37:504.

17p- Syndrome (MDS)

Dobyns, W.B. (1989) The Neurogenetics of Lissencephaly. In Johnson W.G. (editor): *Neurogenetic Diseases.* Neurologic Clinics. Philadelphia. W.B. Saunders Company, pp. 89-105.

17p11.2- Smith-Magenis Syndrome (SMS)

Smith, A.C.M., Mcgavran, L., et al. (1986) Interstitial Deletion of (17)(p11.2p11.2) in Nine patients. *AJMG* 24:393-414.

Greenberg, F., Guzetta, V., et al. (1991) Molecular Analysis of the Smith-Magenis Syndrome: A Possible Contiguous-Gene Syndrome Associated with the del(17) (p11.2). *AJHG* 49:1207-18.

Elsea, S.H., Juyal, R.C., et al. (1995) Haploinsufficiency of Cytosolic Serine Hydroxymethyl transferase in the Smith-Magenis Syndrome. *AJHG* 57: 1342-50.

18q- Syndrome

Krag-Olsen, B., Brask, B.H., Jacobsen, P., Nielsen, J. (1981) "Is There an Increased Risk of Psychosis in Patients with Ring 18 and Deletion Long Arm 18 ?" In Schmid, W. & Nielsen, J. (editors) *Human Behavior and Genetics.* Amsterdam. Elsevier, pp. 211-220.

Silverman, G.A., Schneider, S.S., et al. (1995) The 18q- syndrome: Analysis of Chromosomes by Bivariate Flow Karyotyping and the PCR reveals a

Successive set of Deletion Breakpoints within 18q21.2-q22.2. *AJHG* 56: 926-937.

22q11- VCF Syndrome

Shprintzen, R.J., Goldberg, R.B., et al. (1992) Late Onset Psychosis in the Velo-Cardio-Facial Syndrome. *AJMG* 43:141-142.

Pulver, A.E., Goldberg, R. et al. (1994) Psychotic Illness in Patients Diagnosed with Velo-Cardio-Facial Syndrome and their Relatives. *J Nerv Ment Dis* 182:2887-90.

Linday, E.A., Morris, M.A., et al. (1995) Schizophrenia and Chromosomal Deletions Within 22q11.2. *AJHG* 56:1502-3.

Duplication Syndromes

McDermid, H.E., Duncan, A.M.V., et al. (1986) Characterization of the Supernumary Chromosome in Cat Eye Syndrome. *Science* 232:646-48.

Conclusion

Flint, J, et al. (1995) The Detection of Subtelomeric Chromosome Rearrangements in Idiopathic Mental Retardation. *Nat Genet* 9:132-139.

Ledbetter, D.H. (1995) Minireview: Cryptic Translocations and Telomere Integrity. *AJHG* 51:451-56.

Xq 28- Syndrome

Gedeon, A.K., Keinanen, M., et al. (1995) Overlapping Submicroscopic Deletions in Xq28 in Two Unrelated Boys with Developmental Disorders: Identification of a Gene Near FRAXE. *AJHG* 56:907-914

Behavioral Consequences of Duplication Syndromes

Bartsch, O.,Kalbe, U., et al. (1990) Clinical Diagnosis of Partial Duplication 7q. *AJMG* 37:254-257.

CHAPTER FOUR:
THE BEHAVIORAL CONSEQUENCES OF SPECIFIC SINGLE GENE DISORDERS

Huntington Disease

Gusella, J.,Wexler, N., et al. (1983) A Polymorphic DNA Marker Genetically Linked to Huntingtons Disease. *Nature* 306:234-238.

Caine, E.D., Shoulson, I. (1983) Psychiatric Syndromes in Huntingtons Disease. *Am J Psych* 140: 728.

HD Collaborative Research Group (1993) A Novel Gene Containing a Trinucleotide Repeat That is Expanded and Unstable on Huntington's Disease Chromosomes. *Cell* 72:971-983.

Ranen, N.G., Stine, O.C., et al. (1995) Anticipation and Instability of IT-15 (CAG)n Repeats in Parent-Offspring Pairs with Huntington Disease. *AJHG* 57:593-602.

PKU

Zschocke, J., Graham, C.A., et al. (1995) Phenylketonuria Mutation Analysis in Northern Ireland: A Rapid Stepwise Approach. *AJHG* 57:1311-17.

Dentatorubral-Pallidoluysian Atrophy

Onodera, O., Oyake, M., et al. (1995) Molecular Cloning of a Full-Length cDNA for Dentatorubral-Pallidoluysian Atrophy and Regional Expressions of the Expanded Alleles in the CNS. *AJHG* 57:1050-60.

Wilson Disease

Wilson, S.A.K. (1912) Progressive Lenticular Degeneration: A Family Nervous Disease Associated with Cirrhosis of the Liver. *Brain* 34:295.

Frydmann, M., Bonné-Tamir, B., et al. (1985) Assignment of the Gene for Wilson disease to Chromosome 13. *PNAS* 82:1819-1821

Medalia, A., Scheinberg, I.H. (1989) Psychopathology in Patients with Wilson's disease. *Am J Psych* 146: 662.

Cox, D.W. (1995) Genes of the Copper Pathway. *AJHG* 56: 828-834.

Neurofibromatosis Type 1

Listernick, R., Charrow, J. (1990) Neurofibromatosis type 1 in Childhood. *Journal of Pediatrics* 116: 845-853.
Eldridge, R., Denckla, M., et al. (1989) Neurofibromatosis Type 1 (Recklinghausen's Disease): Neurologic and Cognitive Assessment with Siblings Controls. *Am J Dis Child* 143:833-7.
Eliason, M.J. (1986) Neurofibromatosis: Implications for Learning and Behavior. *Dev Behav Pediatr* 7:175-9.
Varnhagen, C.K., Lewin, S., et al. (1988) Neurofibromatosis and Psychological Processes. *Dev Behav Pediatr* 9:257-65.
Rosman, N.P., Pearce, J. (1967) The Brain in Multiple Neurofibromatosis (von Recklinghausen's Disease): a Suggested Neuropathological Basis for the Associated Mental Defect. *Brain* 90: 829-38.

DDPA

Wilhelmsen, K.C., Lynch, T., et al. (1994) Localization of Disinhibition-Dementia-Parkinsonism-Amyotropy Complex to 17q21-22. *AJHG* 55:1159-65.

Sjögren-Larsson Syndrome (SLS)

Rogers, G.R., Rizzo, W.B., et al. (1995) Genetic Homogeneity in Sjögren-Larsson Syndrome: Linkage to Chromosome 17p in Families of Different Non-Swedish Ethnic Origin. *AJHG* 57:1123-29.

Myotonic Muscular Dystrophy

Alberts, M.J., Roses, A.D. (1989) Myotonic Muscular Dystrophy. In Johnson, W.G. (editor): *Neurogenetic Diseases.* Neurologic Clinics. Philadelphia. W.B. Saunders Company, pp. 1-8.
Willems, et al. (1992) Further Localization of X-linked Hydrocephalus in the Chromosomal Region Xq28. *AJHG* 51:307-315.

X-Linked Mental Retardation Syndromes

Oculocerebrorenal Syndrome

Olivos-Glander, I.M., Jänne, P.A., et al. (1995) The Oculocerebrorenal Syndrome Gene Product is a 105-kD Protein Localized to the Golgi Complex. *AJHG* 57:817-823.

Kenworthy, L., Park, T., et al. (1993) Cognitive and Behavioral Profile of the Oculocerebrorenal Syndrome of Lowe. *AJMG* 46: 297-303.

Haspeslagh, M., et al. (1984) The Coffin-Lowry Syndrome. *Eur. J. Pediatr.* 143:82.

Opitz, J.M., et al. (1982) The FG Syndrome- Further Studies on Three Affected Individuals From the FG Family. *AJMG* 12:147.

Chelly, J., Tumer, Z., et al. (1993) Isolation of a Candidate Gene for Menkes disease that Encodes a Potential Heavy Metal Binding Protein. *Nat Genet* 3:14-19.

Jonsson, J.J., Aronovich, E.L., et al. (1995) Molecular Diagnosis of Mucopolysaccaridosis Type II (Hunter Syndrome) by Automated Sequencing and Computer Assisted Interpretation: Toward Mutation Mapping of the Iduronate-2-sulfatase Gene. *AJHG* 56:597-607.

Lesch, M., Nyhan, W.L. (1964) A Familial Disorder of Uric Acid Metabolism and Central Nervous System Function. *Am. J. Med.* 36:561.

Rosenburg, R.N., Grossman, A. (1989) Hereditary Ataxia. In Johnson, W.G. (editor): *Neurogenetic Diseases.* Neurologic Clinics. Philadelphia. W.B. Saunders Company, pp. 25-34.

Farlow, M.R.,DeMyer, W., Dlouhy, S.R.,Hodes, M.E. (1987) X-linked Recessive Inheritance of Ataxia and Adult-onset Dementia. *Neurology* 37:602-07.

Fragile X

Barnes, D.M. (1989) "Fragile X" Syndrome and Its Puzzling Genetics. *Science* 243:171-72.

Brown, W.T. (1989) The Fragile X Syndrome. In Johnson, W.G. (editor): *Neurogenetic Diseases.* Neurologic Clinics. Philadelphia. W.B. Saunders Company, pp. 107-122.

Wöhrle D, et al. (1992) A Microdeletion of less than 250kb, Including the Proximal Part of the FMR-1 gene and the Fragile-X site, in a Male with the Clinical Phenotype of Fragile-X Syndrome. *AJHG* 51:299-306.

Ashley, A.E., Sherman, S.L. (1995) Population Dynamics of a Meiotic/Mitotic Expansion Model of the Fragile X Syndrome. *AJHG* 57:1414-425.

Loesch, D.Z., Huggins, R., et al. (1995) Expansion of the CGG repeat in

Fragile X in the FMR1 Gene Depends on the Sex of the Offspring. *AJHG* 57:1408-13.
Sherman, S.L. (1995) The High Prevalence of Fragile X Premutation Carrier Females: Is this Frequency Unique to the French Canadian Population ? *AJHG* 57:991-3.

X-Linked Hydrocephalus

Jouet, M., Moncla, A., et al. (1995) New Domains of Neural Cell-Adhesion Molecule L1 Implicated in X Linked Hydrocephalus and MASA Syndrome. *AJHG* 56:1304-14.

Androgen Insensitivity

Schmitt-New, M., Thiele, H., et al. (1995) Two Novel SRY Missense Mutations Reducing DNA binding Identified in XY Females and their Mosiac Fathers. *AJHG* 56:862-869.

Single Gene Disorders with Behavioral Implications That are Not Yet Localized

Rett Syndrome

Rett, A. (1966) Uber ein Eigenartiges Hirnatrophisches Syndrome bei Hyperammonamie im Kindesalter. *Wein Med Wochenschrift* 116:723-6.
Migeon, B.R., Dunn, M.A. (1995) Studies of X-inactivation and Isodisomy in Twins Provide Further Evidence that the X chromosome is not Involved in Rett Syndrome. *AJHG* 56:647-653.

Hallervorden-Spatz

Morphy, M.A., Feldman, J.A., Kilburn, G. (1989) Hallervorden-Spatz Disease in a Psychiatric Setting. *J Clin Psych* 50:66-7.

Androgen Insensitivity

Haqq, C.M., et al. (1994) Molecular Basis of Mammalian Sexual Determination: Activation of Müllerian Inhibiting Substance Gene Expression by SRY. *Science* 266:1494-1500.

Smith-Lemli-Opitz

Alley, T.L., Gray, B.A., et al. (1995) Identification of a Yeast Artificial Chromosome Clone Spanning a Translocation Breakpoint at 7q32.1 in a Smith-Lemli-Opitz Syndrome Patient. *AJHG* 56:1411-1416.

CHAPTER FIVE: APPROACHES TO STUDYING THE GENETICS OF BROAD BEHAVIORAL PHENOTYPES

The Role of Family Studies in Studying Broad Behavioral Phenotypes

Rimmer, J., Chambers, D.S. (1969) Alcoholism: Methadologic Consideration in the Study of Family Illnesses. *Am J Orthopsychiatry* 39: 760-768.

Risch, N. (1983) Estimating Morbidity Risks in Relatives: the Effect of Reduced Fertility. *Behav Genet* 13: 441-451.

Tsuang, M.T., Flemming, J.A., Kendler, K.S.,Gruenberg, A.S. (1988) Selection of Controls for Family Studies. *Arch Gen Psych* 45:1006-1008.

The Role of Twin Studies in Studying Broad Behavioral Phenotypes

Corboz, R.J., Schaub, C. (1981) "Behavior Disturbances fostered by the Twin Situation." In Schmid, W. & Nielsen, J. (editors) *Human Behavior and Genetics*. Amsterdam. Elsevier, pp. 225-230.

Gottesman, I.I., Carey, G. (1983) Extracting Meaning and Direction from Twin Data. *Psych Dev* 1:35-50.

Holzinger, K.J. (1929) The Relative Effect of Nature and Nurture Influences on Twin Differences. *J Educ Psych* 20:241.

Kendler, K. (1983) Overview: Current perspective on Twin Studies of Schizophrenia. *Am J Psychiatry* 140:1413-1425.

Smith, C. (1974) Concordance in Twins: Methods and Interpretation. *Am J Hum Genet* 26:454-466.

Vandenberg, S.G. (1976) "Twin Studies" In Kaplan, A.R. (editor): *Human Behavior Genetics*. Springfield, Illinois. Charles C. Thomas, pp. 90-150.

Neale, M.C., Kendler, K.S. (1995) Models of Comorbidity for Multifactorial Disorders. *AJHG* 57: 935-953.

Segregation Analysis

Lalouel, J.M., Rao, D.C., Morton, D.C., Elston, R.C. (1983) A Unified Model for Complex Segregation Analysis. *Am J Hum Genet* 35:816-26.
Mcgue, M.,Gottesman, I.I., Rao, D.C. (1985) Resolving Genetic Models for the Transmission of Schizophrenia. *Genet Epidemiol* 2: 99-110.
Risch, N.,Baron, M. (1985) Segregation Analysis of Schizophrenia and Related Disorders. *Am J Hum Genetics* 36: 1039-59.
Smith, C.A.B. (1975) A Non-parametric Test for Linkage with a Quantitative Character. *Am J Hum Genet* 38: 451-60.

CHAPTER SIX: THE GENETICS OF BROAD BEHAVIORAL PHENOTYPES

The Genetics of Mental Retardation

Camarino, G.,Mattei, M.G., Mattei, J.F., Jaye, M., Mandel, J.L. (1983) Close Linkage of Fragile X Mental Retardation Syndrome to Haemophilia B and Transmission Through a Normal Male. *Nature* 306: 701-704.
Farlow, M,R,,DeMyer, W.,Diouhym, S.R., Hodes, M.E. (1987) X-Linked Recessive Inheritance of Ataxia and Adult-onset Dementia: Clinical Features and Preliminary Linkage Analysis. *Neurology* 37(4): 602-7.
Fryns, J.P. (1986) The Female and the Fragile X: A Study of 144 Obligate Female Carriers. *Am J of Medical Genetics* 23:157-169.

The Genetics of Schizophrenia

Major Reviews of the Genetics of Schizophrenia

Kallmann, F.J. (1938) *The Genetics of Schizophrenia.* New York. Augustin.
Rosenthal, D., Kety, S.S. (1968) *The Transmission of Schizophrenia.* Oxford. Pergamon.
Gottesman, I.I.,Shields, J. (1972) *Schizophrenia and Genetics: A Twin Study Vantage Point.* New York. Academic Press.
Liston, E.H., Jarvik, L.F. (1976) "Genetics of Schizophrenia." In *Genetics and Psychiatry.* New York. Basic Books, pp. 76-94.
Gottesman, I.I.,Shields, J. (1982) *Schizophrenia: The Epigenetic Puzzle.* New York. Cambridge University Press.
Kendler, K.S. (1986) "Genetics of Schizophrenia" In Frances, A.J. and Hales, R.E. (editors) *Annual Review Vol. 5.* Washington: American Psychiatric Press, pp. 25-41.

Rudin, E. (1916) *Zur Verebung und Neuentstehung der Dementia Praecox.* Berlin. Springer.

Kahn, E. (1923) "Studien über Vererbüng und Enstehung geistiger Störungen." In Vol IV, *Schizoid und Schizophrenie in Erbgang.* Berlin: Springer.

Schulz, B. (1927) Zur Frage einer Belastunsstatistik der Durchschnittsbevölkerung. Geschwisterschaften und Elternschaften von 100 Hirnarteriosklerotiker-Ehegatten. *Z. ges Neurol. Psychiatr* 109:15-48.

Luxenburger, H. (1928) Vorläufiger Bericht über psychiatrische Serienuntersuchungen an Zwillingen. *Z. ges Neurol. Psychiatr* 116:297-326.

Brugger, P. (1929) Zur Frage einer Belastungsstatistik der Durchschnittsbevölkerung. *Z. ges Neurol. Psychiatr* 118:459-488.

Schulz, B. (1932) Zur Erbpathologie der Schizophrenie. *Zeitschrift für die gesamte Neurologie und Psychiatre* 143:175-293.

Klemperer, J. (1933) Zur Belastungsstatistik der Durchschnittbevölkerung: Psychosenhäufigkeit unter 1000 stichprobenmässig aus den Geburtsgegistern der Stadt München (Jahring 1881-1890) ausgelesen Probanden. *Z. ges Neurol. Psychiatr.* 146: 277-316.

Sjögren, T. (1935) Investigations of the Heredity of Psychoses and Mental Deficiency in Two North Swedish Parishes. *Ann Eugen. Camb* 6: 253-318.

Stömgren, E. (1935) Zum Ersatz des Weinbergschen abgekürtzen Verfahrens, Zugleich ein Beitrag zur Frage von der Erblichkeit des Erkrankungsalters bei der Schizophrenie. *Z ges Neurol. Psychiatr* 152:784-797.

Kallmann, F.J. (1938) *The Genetics of Schizophrenia.* New York. Augustin.

Schulz, B. (1940) Kinder Schizophrener Elternparre. *Zeitschrift für die gesamte Neurologie und Psychiatrie.* 168: 332-381.

Fremming, K. (1951) *The Expectation of Mental Infirmity in a Sample of the Danish Population.* Occassional Papers in Eugenics, No. 7. London: Cassell and Co.

Elsässer, G. (1952) *Die Nachkommen geisteskranker Eltenparre.* Stuttgart. Thieme.

Lewis, A.J. (1957) The Offspring of Parents both Mentally Ill. *Acta Genetica* (Basel) 7:349-365.

Reisby, N. (1967) Psychosis in Children of Schizophrenic Mothers. *Acta Psych. Scand.* 43:8-30.

Zerbin-Rudin, E. (1967) "Endogene Psychosen Schizophrenien." In Becker, P.E.

(Èditor) *Humangenetik*. Stuttgart. Theime, pp. 446-504.

Erlenmeyer-Kimling, L. (1968) Studies on the Offspring of Two Schizophrenic Parents. In Rosenthal, D. & Kety, S. (editors) *The Transmission of Schizophrenia*. Oxford. Pergamon Press.

Erlenmeyer-Kimling, L. (1968) Studies on the Offspring of Two Schizophrenic Parents. *J. Psych. Res*. 6:65-83.

Tsuang, M.T.,Winokur, G.,Crowe, R.R. (1980) Morbidity risks of Schizophrenia and Affective Disorders Among First Degree Relatives of Patients with Schizophrenia, Mania, Depression, and Surgical Conditions. *Br J Psych* 137:497-504.

Scharfetter, C.,Nusperli, M. (1980) The Group of Schizophrenias, Schizoaffective Psychoses and Affective Disorders. *Schizophr Bull* 6:586-591.

Bleuler, M. (1981) "Human Behavior and Genetic Research, Views of a Psychiatrist." In Schmid W. & Nielsen, J. (editors) *Human Behavior and Genetics*. Amsterdam. Elsevier, pp. 5-8.

Pope, H.G., Jones, J.M., Cohen, B.M., et al. (1982) Failure to Find Evidence of Schizophrenia in First Degree Relatives of Schizophrenic Probands. *Am J Psych* 139:826-828.

Guze, S.B., Cloninger, C.R.,Martin, R.L., et al. (1983) A Follow-up and Family Study of Schizophrenia. *Arch Gen Psych* 40: 1273-1276.

Karlsson, J.L. (1983) Family Transmission of Schizophrenia: a Review and Synthesis. *Br J Psych* 140:600-606.

Baron, M., Gruen, R., Asnis, L., et al. (1985) A Family Study of Schizophrenia and Normal Control Probands: Implications for the Spectrum Concept of Schizophrenia. *Am J Psych* 142:447-455.

Kendler, K.S., Gruenberg, A.M., Tsuang, M.T. (1985) Psychiatric Illness in the First-Degree Relatives of Schizophrenic and Surgical Control Patients: a Family Study Using DSM III Criteria. *Arch Gen Psych* 42: 770-779.

Alda, M., Dvorakova, M.,Zvolsky, P., Papezova, H. (1988) A Family Study of Chronic Schizophrenia. *Schiz Res* 1:126-7.

Karlsson, J.L. (1988) Partly Dominant Transmission of Schizophrenia in Iceland. *Brit J Psych* 152:324-329.

Twin Studies in Schizophrenia

Luxenburger, H. (1928) Vorläufiger Bericht über psychiatrishe Serienuntersuchungen an Zwillingen. *Zeitschrift Für die gesamte Neurologie und Psychiatrie* 116:297-326.

Luxenburger, H. (1930) Psychiatrisch-neurologische Zwillingspathologie. *Zerblatt für die Gesamte Neurologie und Psychiatrie* 56:145-180.

Luxenburger, H. (1934) Die Manifestionswarhschleinlichkeit der Schizophrenie im Licht der Zwillingsforschung. *Zeitschrift fur Psychische Hygeine* 7:174-184.

Rosanoff, A.J.,Handy, L.M.,Plesset, I.R.,Brush, S. (1934) The Etiology of so-called Schizophrenic Psychoses with Special Reference to their Occurrence in Twins. *Am J Psych* 91:247-286.

Luxenburger, H. (1936) Untersuchungen an Schizophrenen Zwillingen und ihren Geschwistern zur Prüfung der Realitat von Manifestationsschwankungen. *Zeitschrift für die gesamte Neurologie und Psychiatrie* 154:351-394.

Essen-Möller, E. (1941) Psychiatrische Untersuchungen an einer Serie von Zwillingen. *Acta Psychiatrica et Neurologica Scandinavia* Supplement 3.

Kallmann, F.J. (1946) The Genetic Theory of Schizophrenia: an Analysis of 691 Schizophrenic Twin Index Families. *Am J Psych* 103: 309-322.

Slater, E., Shields, J. (1953) *Psychotic and Neurotic Illnesses in Twins*. Med Res Council Spec Rep Ser 278, London: Her Majesty's Stationary Office.

Inouye, E. (1963) Similarity and Dissimilarity of Schizophrenia in Twins. *Proceedings of the 3rd World Congress of Psychiatry*. Montreal. University of Toronto Press, pp. 524-530.

Tienari, P. (1963) Psychiatric Illnesses in Identical Twins. *Acta Psychiatr Scand* Suppl 171.

Kringlen, E. (1964) *Schizophrenia in Male Monozygotic Twins*. Oslo: University Press. Also in *Acta Psychiatr Scand* 40 (suppl 178) 1-76.

Harvald, B., Hauge, M. (1965) "Hereditary Factors Elucidated by Twin Studies." In Neel, J.V. et al. (editors) *Genetics and the Epidemiology of Chronic Diseases*. Washington. Dept. of Health Education and Welfare, pp. 61-76.

Kringlen, E. (1966) Schizophrenia in Twins, an Epidemiological-clinical Study. *Psychiatry* 29: 172-184.

Gottesman, I.I., Shields, J. (1966) Schizophrenia in Twins: a 16 years' Consecutive Admissions to a Psychiatric Clinic. *Br. J. Psych*. 112: 809-818.

Kringlen, E. (1967) *Heredity and Environment in the Functional Psychoses: An Epidemiological-clinical Twin Study*. Oslo. Universitetsforlaget. or London. Heineman.

Tienari, P. (1968) Schizophrenia in Monozygotic Male Twins. In Rosenthal, D. & Kety, S.S. (editors) *The Transmission of Schizophrenia* Oxford. Pergamon, pp. 27-36.

Kringlen, E. (1968) An Epidemiologic - clinical Twin Study on Schizophrenia.

In Rosenthal, D. & Kety, S.S. (editors) *Transmission of Schizophrenia.* Oxford. Pergamon, pp. 49-63.

Fisher, M., Harvald, B., Hauge, M. (1969) A Danish Twin Study of Schizophrenia. *Br J of Psych* 115:981-990.

Pollin, W., Allen, M.G., Hoffer, A., Stabenau, J.R., Hrubec, Z. (1969) Psychopathology in 15,909 Pairs of Veteran Twins: Evidence for a Genetic Factor in the Pathogenesis of Schizophrenia and its Relative Absence in Psychoneurosis. *Am J Psychiatry* 126:597-610.

Essen-Möller, E. (1970) Twenty-one Psychiatric Cases and their MZ co-twins: A Thirty Year's Follow-up. *Acta Geneticae Medicae et Gemellologiae* 19:315-317.

Tienari, P. (1971) Schizophrenia and Monozygotic Twins. In Achté, K.A. (editor.) *Psychiatria Fennica,* Helsinki. Helsinki University Central Hospital, pp. 97-104.

Fisher, M. (1971) Psychosis in the Offspring of Schizophrenic Monozygotic Twins and their Normal Co-Twins. *Br J Psych* 118:43-52.

Inouye, E. (1972) "A Search for a Research Framework of Schizophrenia in Twins and Chromosomes." In Kaplan, A.R. (editor) *Genetic Factors in Schizophrenia.* Springfield. Thomas. pp. 495-503.

Mitsuda, H. (1972) "Heterogeneity of Schizophrenia." In Kaplan, A.R. (editor) *Genetic Factors in Schizophrenia.* Springfield. Thomas. pp. 276-293.

Allen, M.G., Cohen, S., Pollin, W. (1972) Schizophrenia in Veteran Twins: a Diagnostic Review. *Am J Psych* 128:939-945.

Gottesman, I.I., Shields, J. (1972) *Schizophrenia and Genetics: A Twin Study Vantage Point.* New York. Academic Press.

Tienari, P. (1975) Schizophrenia in Finnish Male Twins. In *Studies of Schizophrenia.* (editor) Lader, M.H. Ashford, England. Headley Brothers.

Kendler, K., Robinette, C.D. (1983) Schizophrenia in the NAS-NRC Twin Registry: a 16 Year Update. *Am J Psych* 140:155-156.

McGuffin, P., et al. (1984) Twin Concordance for Operationally Defined Schizophrenia: Confirmation of Familiarity and Heritability. *Arch Gen Psych* 41:541-45.

Conclusions from Schizophrenia Twin Studies

Essen-Möller, E. (1941) Empirische Ähnlichkeitsdiagnose bei Zwillingen. *Hereditas, Genetiskt Arkiv* 27:1-50.

Rosenthal, D. (1959) Some Factors Associated with Concordance and Discordance with Respect to Schizophrenia in Monozygotic Twins, *J Nerv*

Ment Dis 129:1-10.
Rosenthal, D. (1962) Problems of Sampling and diagnosis in the Major Twin Studies of Schizophrenia. *J Psych Res* 1: 116-134.
Erlenmeyer-Kimling, L., Paradowski, W. (1966) Selection and Schizophrenia. *Am Naturalist* 100: 651.
Shields, J., Gottesman, I.I., Slater, E. (1967) Kallmann's 1946 Twin Study in the Light of New Information. *Acta Psychiatrica Scandanavia* 43:385-396.
Gottesman, I.I. (1968) Severity Concordance in the Maudsley-Bethlam Schizophrenia Twin Study. In Rosenthal, D. (editor) *The Transmission of Schizophrenia*. Oxford. Pergamon Press.
Hoffer, A., Pollin, W. (1970) Schizophrenia in the NAS-NRC panel of 15,909 Veteran Twin Pairs. *Arch Gen Psych* 23:469-477.
Inouye, E. (1972) Monozygotic Twins with Schizophrenia Reared Apart in Infancy. *Jap. J. Hum. Genet.* 16:182-190.
Fisher, M. (1973) Genetic and Environmental Factors in Schizophrenia. *Acta Psych Scand* [Suppl] 238:1-152.
Gottesman, I.I., Shields, J. (1976) A Critical Review of Recent Adoption, Twin and Family Studies of Schizophrenia: Behavioral and Genetic Perspectives. *Schizophrenia Bull* 2:360.
Shields, J. (1982) *Schizophrenia: The Epigenetic Puzzle*. Cambridge. Cambridge University Press.
Kendler, K. (1983) Overview: Current Perspective on Twin Studies of Schizophrenia. *Am J Psychiatry* 140:1413-1425.
Farmer, A.E., Gottesman, I.I. (1984) Searching for the Split in Schizophrenia: a Twin Perspective. *Psych Research* 13:109 118.
Dworkin, R.H., Lenzenweger, M.F., Moldin, S.O., Skillings, G.F., Levick, S.E. (1988) A Multidimensional Approach to the Genetics of Schizophrenia. *Am J Psych* 145:1077-83.

Adoption Studies in Schizophrenia

Heston, L.L. (1966) Psychiatric Disorders in Foster Home Reared Children of Schizophrenic Mothers. *Br J Psych* 112: 819.
Karlsson, J.L. (1966) *The Biological Basis of Schizophrenia*. Springfield, Illinois. C.C. Thomas.
Kety, S.S., Rosenthal, D., Wender, P.H., Schulsinger, F. (1968) The Types and Prevalence of Mental Illness in the Biological and Adoptive Families of Adopted Schizophrenics. In *The Transmission of Schizophrenia*. Oxford. Pergamon.

Kety, S.S., Rosenthal, D., Wender, P.H., et al. (1968) The Types and Prevalence of Mental Illness in the Biological and Adoptive Families of Schizophrenics. *J Psych Res* 6:345-362.

Rosenthal, D.,Wender, P.H., Kety, S.S., Schlusinger, F., Welner, J., Ostergaard, L. (1968) Schizophrenics Offspring Reared in Adoptive Homes. In *Transmission of Schizophrenia* (editors) Rosenthal, D. and Kety, S.S. Oxford. Pergamon, pp. 377-91

Wender, P.H., Rosenthal, D., Kety, S.S. (1968) A Psychiatric Assessment of the Adoptive Parents of Schizophrenics. In *Transmission of Schizophrenia* eds Rosenthal, D. and Kety, S.S. Oxford. Pergamon.

Kety, S.S., Rosenthal, D.,Wender, P.H. et al. (1971) Mental Illness in the Biological and Adoptive Families of Adoptive Schizophrenics. *Am J Psych* 138: 302-306.

Rosenthal, D., Wender, P.H., Kety, S.S., et al. (1971) The Adopted Away Offspring of Schizophrenics. *Am J Psych* 128: 307-311.

Wender, P.H., Rosenthal, D., Zahn, T.P., et al. (1971) The Psychiatric Adjustment of the Adopting Parents of Schizophrenics. *Am. J. Psych.* 127:1013-1018.

Wender, P.H., Rosenthal, D., Kety, S.S., et al. (1974) Crossfostering: a Research Strategy for Clarifying the Role of Genetic and Experimental Factors in the Etiology of Schizophrenia. *Arch Gen Psych* 30:121-128.

Kety, S.S., Rosenthal, D., Wender, P.H., et al (1975) Mental Illness in the Biological and Adoptive Families of Adopted Individuals Who Became Schizophrenic: a Preliminary Report Based on Psychiatric Interviews. In *Genetic Research in Psychiatry*. (editors) Fieve, R.R.,Rosenthal, D., Brill, H.H. Baltimore. Johns Hopkins.

Higgins, J. (1976) Effects of Child Raring by Schizophrenic Mothers : a Follow-up. *J Psychiatr Res* 13:1-9.

Wender, P.H., Rosenthal, D., Ranier, J.D., et al. (1977) Schizophrenics' Adopting Parents: Psychiatric Status. *Arch General Psych* 34:777-784.

Kety, S.S., Rosenthal, D., Wender, P.H., et al. (1978) The Biologic and Adoptive Families of Adopted Individuals who Became Schizophrenic: Prevalence of Mental Illness and other Characteristics. In *The Nature of Schizophrenia*. (editor) Wynn, L.C. New York. John Wiley.

Lowing, P.A.,Mirsky, A.F.,Pereira, R. (1983) The Inheritance of Schizophrenic Spectrum Disorders: a Reanalysis of the Danish Adoptee Study Data. *Am J Psych* 140:1167-1171.

Rudin, E. (1916) *Zur Verebung und Neuentstehung der Dementia Praecox.* Berlin. Springer.
Book, J. (1953) A Genetic and Neuropsychiatric Investigation of a North Swedish Population. *Acta Genet Statist Med* 4:1.
Book, J. (1953) Schizophrenia as a Gene Mutation. *Acta Genet Statist Med* 4:133.
Roth, M. (1957) "Interaction of Genetic and Environmental Factors in the Causation of Schizophrenia." In Richter, D. (editor) *Schizophrenia, Somatic Aspects*. New York. Macmillan.
Slater, E. (1958) A Monogenic Theory of Schizophrenia. *Acta Genet Statist Med* 8:50.
Edwards, J.H. (1960) The Simulation of Mendelism. *Acta Genet Statist Med* 10:63.
Garrone, G. (1962) Étude statistique et génétique de la schizophrenie a Genéve de 1901 a 1950. *J Genet Hum* 7:189.
Karlsson, J.L. (1964) A Hereditary Mechanism for Schizophrenia Based on Two Separate Genes, one Dominant, the other Recessive. *Hereditas* 51:74.
Gottesman, I.I., Shields, J. (1967) A Polygenic Theory of Schizophrenia. *Proc Nat Acad Science* 58:199-205.
Heston, L.L. (1970) The Genetics of Schizophrenic and Schizoid Disease. *Science* 167:249.
Elston, R.C., Campbell, M.A. (1970) Schizophrenia: Evidence for the Major Gene Hypothesis. *Behav Genetic* 1:3.
Kidd, K.K., Cavilli-Sforza, L.L. (1973) An Analysis of the Genetics of Schizophrenia . *Soc Biol* 20:254.
Eberhard, G., et al. (1975) Schizophrenia Susceptability and HLA Antigens. *Neuropsychobiology* 1: 211-17.
Book, J.A.,Wetterberg, L.,Modrzwewski, K. (1977) Schizophrenia in a North Swedish Geographic Isolate, 1900-1977. Epidemiology, Genetics and Biochemistry.
Elston, R.C., Namboodiri, K.K., Spence, M.A., et al. (1978) A Genetic Study of Schizophrenia Pedigrees, II: One Locus Hypothesis. *Neuropsychobiology* 4:193-206.
Carter, C.L., Chung, C.S. (1980) Segregation Analysis of Schizophrenia Under a Mixed Genetic Model. *Hum Hered* 30:350-356.
Crow, T.J. (1980) Molecular Pathology of Schizophrenia: More Than Just One Disease Process. *Brit Med J* 280:66-68.
Rao, D.C., Morton, N.E., Gottesman, I.I., Low, R. (1981) Path Analysis of

Qualitative Data on Pairs of Relatives: Application to Schizophrenia. *Hum Hered* 31:325-33.

Baron, M. (1982) Genetic Models of Schizophrenia. *Psych Scand* 65:263-75.

Tsuang, M.T., Bucher, K.D., Fleming, J.A (1982) Testing the Monogenic Theory of Schizophrenia: an Application of Segregation Analyisis to Blind Family Study Data. *Br J Psych* 140:595-599.

Orourke, D.H.,Gottesman, I.I,. et al. (1982) Refutation of the Single Locus Model in the Etiology of Schizophrenia. *Amer J of Human Genetics* 34:630-649.

Farmer, A.E., McGuffin, P., Spitznagel, E. (1983) Heterogeneity in Schizophrenia : a Cluster Analytic Approach. *Psych Res* 8:1-12.

Dworkin, R.H., Lenzenweger, M.F. (1984) Symptoms and the Genetics of Schizophrenia. Implications for Diagnosis. *Amer J of Psych* 141.1541-46.

Mirsky, A.F., Duncan-Johnson, C.C. (1984) Nature vs Nurture in Schizophrenia: the Struggle Continues. *Integrative Psychiatry* 2:137-141.

Crow TJ (1985) The Two System Concept: Origin and Current Status. *Schizophrenia Bulliten* 11:471-86.

Risch, N.,Baron, M. (1985) Segregation Analysis of Schizophrenia and Related Disorders. *Am J Hum Genetics* 36: 1039-59.

Mcgue, M.,Gottesman, I.I., Rao, D.C. (1985) Resolving Genetic Models for the Transmission of Schizophrenia. *Genet Epidemiol* 2: 99-110.

Eaves, L.J., Kendler, K.S., et al. (1986) The Familial Sporadic Classification, It's Power for the Resolution of Genetic and Environmental Etiologic Factors. *Jour Psych Res* 20: 115-130.

McGuffin, P.,Farmer, A.,Gottesman, I.I. (1987): Is there Really a Split in Schizophrenia? the Genetic Evidence. *Brit Journal of Psychiatry* 150:581-592.

Holzman, P.S.,Kringlen, E.,Matthysse, S.,et al. (1988) A Single Dominant Gene can Account for Eye Tracking Dysfucnctions and Schizophrenia in Offspring of Discordant Twins. *Arch Gen Psych* 45:641-47.

The Genetics of Bipolar Disorder

Major Reviews of the Genetics of Bipolar Disorder

Kallman, F.J. (1954) "Genetic Principles in Manic Depressive Psychosis" In Zubin, J. and Hoch, R. (editors) *Depression*. New York. Grune & Stratton.

Gershon, E.S.,Bunney, W.E. Jr., Leckman, J.F., et al. (1976) The Inheritance of Affective Disorders: a Review of Data and Hypotheses. *Behav Genet*

6:227-261.
Cadoret, R.J., Winokur, G. (1976) "Genetics of Affective Disorders." In Sperber, M.A. and Jarvik, L.F. (editors) *Psychiatry and Genetics*. New York. Basic Books.
Nurnberger, J.I.,Gershon, E.S. (1982) Genetics of Affective Disorders, In *Handbook of Affective Disorders*. (editor) Paykel, E. London. Churchill Livingstone.
Gershon, E.S., Berrettini, W., Nurnberger, J., Goldin, L.R. (1987) Genetics of Affective Illness. In Meltze, H.Y. (editor) *Psychopharmacology: The Third Generation of Progress*. Raven Press, pp. 481-491.
Blehar, M.C., Weissman, M.M., Gershon, E.S., Hirschfield, R.M.A. (1988) Family and Genetic Studies of Affective Disorders. *Arch Gen Psych* 45:289-92.
Mendlewicz, J. (1988) Population and Family Studies in Depression and Mania. *Brit J Psych* 153 (Suppl. 3) 16-25.

General

Angst, J., Perris, C. (1968) Nosology of Endogenous Depression: a Comparison of the Findings of Two Studies. *Arch Psychiat. Nervenkr.* 210: 373.
Asano, N. (1967) Clinico-Genetic Study of Manic Depressive Psychoses. In Mitsuda. H. (Editor) *Clinical Genetics in Psychiatry.* Kyoto, Japan. Banko-sha Co.
Leonhard, K. (1959) *Aufteilung der Endogenen Psychosen.* Berlin. Akademie-Verlag.
Leonhard, K., Korff, I., Schulz, H. (1962) Die Temperamente in den Familien der monopolaren und bipolaren phasischen Psychosen. *Psychiat. Neurol.* 143:416.
Perris, C. (editor) (1966) A Study of Bipolar (manic-depressive) and Unipolar Recurrent Depressive Psychosis. *Acta Psychiatr. Scand.* 42:1 (Suppl. 194).
Perris, C. (1968) Genetic Transmission of Depressive Psychoses. In Retterstøl, N., Magnussen, F. (Editors) *Acta Psychiat Scand Suppl* 203, 1968.
Zerbin-Rüdin, E. (1967) Endogene Psychosen. In *Human-Genetik: ein Kurzs Handbuch in Fünf Banden* (editor) Becker, P.E. Vol. 2. Stuttgart. Theime.

Family Studies in Bipolar Disorder

Hoffmann, H. (1921) *Die Nachkommenschaft bei endogenen Psychosen.* Monogr. Neurol Psychiatry 26. Berlin. Springer.

Rüdin, E. (1923) Über Vererbung geistiger Storüngen . *Z. Ges Neur Psychiat* 81:459-96.

Banse, J. (1929) Zum Probleme der Erbprognosebestimmung (Die Erkrankungsaussichten der Vettern und Basen von Manisch -Depressiven) *Z. Ges Neurol Psychiat* 119:576-612.

Luxenburger, H. (1930) Psychiatrisch-neurlogische Zwillings-pathologie. *Z. Ges Neurol. Psychiat* 56: 145.

Luxenburger, H. (1932) Erbprognose und praktische Eugenik im cyclothymen Kriese. *Nervenarzt* 5: 505-18.

Weinberg, I., Lobstein, J. (1936) Beitrag zur Vererbung des maisch-depressiven Irreseins. *Psych Neurol Bl (amst)* Ia 339-72, 1936.

Röll, A., Entres, J.L. (1936) Zum problem der Verarbeitung der Erbprognosebestimmung. Die Erkrankungsaussichten der Neffen und Nichten von Manisch-Depressiven. *Z. Neurol* 156: 169-202.

Slater, E. (1938) Zur Erbpathologie des manisch-depressiven Irreseins. Die eltern und Kinder von Manisch-Depressiven. *Zeitshrift für die gesamte Neurologie und Psychiatrie* 163: 1-47.

Stenstedt, A. (1952) A Study in Manic Depressive Psychoses. *Acta Psych et Neurol Scand Suppl* 79, 1-111.

Reich, T., Clayton, P., Winokur, G. (1969) Family History Studies V: The Genetics of Mania. *Am. J. Psych.* 125:1358-1369.

Perris, C. (1971) Abnormality on Paternal and Maternal Sides: Observations in Bipolar (manic -depressive) and Unipolar Depressive Psychosis. *Brit J Psych* 118:207-210.

Slater, E., Maxwell, J., Price, J.S. (1971) Distribution of Ancestral Secondary Cases in Bipolar Affective Disorders. *Brit J Psych* 118:543, 215-218.

Taylor, M., Abrams, R. (1973) Manic States: a Genetic Study of Early and Late Onset Affective Disorders. *Arch Gen. Psych* 28:656-658.

Winokur, G., Crowe, R.R. (1983) Bipolar Illness. The Sex Polarity Effect in Affectively Ill Family Members. *Arch Gen Psych* 40:57-58, 1983.

Weissman, M.M., Gershon, E.S., Kidd, K.K., et al. (1984) Psychiatric Disorders in the Relatives of Probands with Affective Disorders: The Yale University-National Institute of Health Colaborative Study. *Arch Gen Psych* 41:13-21.

Weissman, M.M., Merikangas, K.R.,Wickramaratne, P.,et al. (1986) Understanding the Clinical Heterogeneity of Major Depression Using Family Data. *Arch gen Psych* 43:430-434.

Torgersen, S. (1986) Genetic Factors in Moderaltely Severe and Mild Affective Disorder. *Arch Gen Psych* 43:222-226.

Twin Studies in Bipolar Disorder

Luxenberger, H. (1932) Endogener Schwachsinn und geschlectsgebundener Erbgang, Z. *Ges. Neurol. Psychiat.* 140:320.

Rosanoff, A.J., Handy, L.M., Plesset, I.R. (1935) The Etiology of Manic-depressive Syndromes with Special Reference to their Occurence in Twins. *Amer J Psych* 91:725-62.

Essen-Möller, E. (1941) Psychiatrische untersuchungen an einer serie von zwillingen. *Acta Psych Scand* (suppl) 23:1.

Slater, E. (1953) Psychotic and Neurotic Illnesses in Twins. *Med Res Council Spec Rep Ser* 278.

Kallmann, F.J. (1954) "Genetic Principles in Manic-Depressive Psychoses." In *Depression* Hoch, P. & Zubin, J. (editors) New York. Grune and Stratton.

De Fonseca, A.F. (1959) *Analise heredo-clinica da pertubacoes affectivas.* Dissertation. University of Porto.

Harvald, B., Hauge, M. (1965) "Hereditary Factors Elucidated by Twin Studies." In Neel, J.V. et al. (editors) *Genetics and the Epidemiology of Chronic Diseases* Washington D.C. Dept of Health Educ. and Welfare Publication.

Kringlen, E. (1967) *Heredity and Environment in the Functional Psychoses. An Epidemiological- Clinical Twin Study*. Oslo, Norway. Universitstoforlaget.

Price, J. (1968) "The Genetics of Depressive Behavior." In *Recent developments in Affective Disorders*. Coppen, A. (editor) *Br J Psych* (Special Pub 2).

Bertelsen, A., Harvald, B., Hauge, M. (1977) A Danish Twin Study of Manic-Depressive Disorders. *Br J Psychiatry* 130:330-351.

Baron, M. (1980) Genetic Models of Affective Disorder: Application to Twin Data. *Acta Genet Med Gemellol* 29:289-294.

Torgersen, S. (1986) Genetic Factors in Moderately Severe and Mild Affective Disorders. *Arch Gen Psych* 43:222-226.

Mode of Inheritance in Bipolar Disorder

Rosanoff, A.J., Handy, L.M., Plesset, I.R. (1935) The Etiology of Manic-Depressive Syndromes with Special Reference to their Occurence in Twins. *Amer J Psych* 91:725-62.

Slater, E. (1936) The Inheritance of Manic Depressive Insanity. *Proc Roy Soc Med* 29:981-90.

Lenz, F. (1937) "To Mendelize Mental Illnesses?" (Trans) *Zeitschrift Fur*

Induktive Abstammungs und vererburg lehre. 73: 559-564.
Slater, E. (1938) Zur Erbpathologie des manisch-depressiven Irreseins. Die eltern und Kinder von Manisch-Depressiven. *Zeitshrift für die gesamte Neurologie und Psychiatrie* 163: 1-47.
Kallman, F.J. (1954) "Genetic Principles in Manic Depressive Psychosis" In Zubin, J. and Hoch, R. (editors) *Depression.* New York. Grune & Stratton.
Winokur, G. and Tanna, V.L. (1969) Possible Role of x-linked Dominant Factor in Manic-depressive Disease. *Dis Nerv Syst* 30:89, 1969.
Mendlewicz, J., Ranier, J.D. (1974) Morbidity Risk and Genetic Transmission in Manic-depressive Illness. *Am J Hum Genetics* 25:692-701.
O'Rourke, D.H., McGuffin, P., Reich, T. (1983) Genetic Analysis of Manic Depressive Illness. *Am J Physical Anthropology* 62:51-59.
Hodgkinson, S., Sherrington, R., Gurling, H., et. al. (1987) Molecular Genetic Evidence for Heterogeneity in Manic Depression. 325:805-06.

The Genetics of Unipolar Disorder (Major Depression)

Bynum, W.F., Porter, R., Shepherd, M. (1985) *Anatomy of Madness.* Volume One. New York. Tavistock. The quote is from Francis Reynolds in the chapter, "The Hunger of the Imagination: Approaching Samuel Johnson's Melancholy." Johnson himself has stated the he had " a vile melancholy, inherited from my father." Bates, W.J. (1977) *Life of Johnson.* New York. HBJ.
Winokur, G.,Pitts, F.N. Jr. (1965) Affective Disorders: VI. a Family History Study of Prevalence , Sex Differences and Possible Genetic Factors. *J Psych Res* 3:113-123.
Winokur, G.,Clayton, P. (1967) Family History Studies : I. Two Types of Affective Disorders Separated According to Genetic and Clinical Factors. In *Recent Advances in Biological Psychiatry.* IX. New York. Plenum Press.
Martin, S., Cadoret, R.J., Winokur, G., Ora, E. (1972) Unipolar Depression: a Family History Study. *Biol. Psych.* 4:205-213.
Winokur, G. (1981) *Depression: The Facts.* London. Oxford University Press.
Mendlewicz. J., Baron, M. (1981) Morbidity Risks in Subtypes of Unipolar Depressive Illness; Differences Between Early and Late Onset Forms. *Br J Psych* 139:463-466.

Primary Degenerative Dementia of the Alzheimer Type

Heyman, A., Wilkerson, W.E., Hurwitz, B.J., et al. (1983) Alzheimers Disease:

Genetic Aspects and Associated Clinical Disorders. *Ann Neurol* 14:507-516.
Wright, A.F., Whalley, L.J. (1984) Genetics, Ageing and Dementia *Br J Psych* 145: 20-38.
St. George-Hyslop, P.H., Tanzi, R.E.,Polinsky, et al. (1985) The Genetic Defect Causing Familial Alzheimer's Disease Maps on Chromosome 21. *Science* 235: 885-90.
Warren, A.C., Robakis, N.K.,Ramakrishna, N., et al. (1987) Beta Amyloid Gene Not Present in Three Copies in Alzheimers Disease. *Genomics* 1:307-12.
Tanzi, R.E.,Gusella, J.F.,Watkins, P.C., et al. (1987) Amyloid Beta Protein Gene : cDNA ,mRNA Distribution and Genetic Linkage near the Alzheimers Locus. *Science* 235:880-884.
Small, G.W.,Greenberg, D.A. (1988) Biological Markers,Genetics,and Alzheimers Disease. *Arch Gen Psych* 45:945-947.
Schellenberg, G,D,,Bird, T,D,,Wijsman, E,M,, et al, (1988) Absence of Linkage of Chromosome 21q21 Markers to Familial Alzheimers Disease. *Science* 241: 1507-1510.
Zubenko, G.S.,Huff, F.J.,Beyer, J.,Auerbach, J.,Teply, I. (1988) Familial Risk of Dememtia Associated with a Biological Subtype of Alzheimers Disease. *Arch Gen Psych* 45: 889-900.
Clark, R.F., Goat, A.M. (1993) Molecular Genetics of Alzheimer's Disease. *Arch Neurology* 50:1164-1172.
Rao, V.S., van Duijn, C.M., et al. (1994) Multiple Etiologies for Alzheimer Disease are Revealed by Segregation Analysis. *AJHG* 55:991-1000.
van Duijn, C.M., Hendriks, L., et al. (1994) A Population-Based Study of Familial Alzheimer Disease: Linkage to Chromosome 14,19, and 21. *AJHG* 55:714-27.

The Genetics of Alcoholism

Kaij, L. (1960) *Alcoholism in Twins: Studies on the Etiology and Sequels of Abuse of Alcohol.* Stockholm. Almquist and Wiksell.

Adoption Studies

Goodwin, D.W., Shulsinger, F. et al. (1973) Alcohol Problems in Adoptees raised apart from alcoholic biological parents. *Arch Gen Psych* 28: 238-43.
Goodwin, D.W., Moller, N., et al. (1974) Drinking Problems in Adopted and Non-Adopted Sons of Alcoholics. *Arch Gen Psych* 31:164-9.

Goodwin, D.W., Knop, J., et al. (1977) Alcoholism and Depression in Adopted-Out Daughters of Alcoholics. *Arch Gen Psych* 34:751-5.

The Genetics of Stuttering

Makuen, G.H. (1914) A Study of 1,000 Cases of Stammering with Special Reference to the Etiology and Treatment of the Affliction. *Therapeutic Gazette* 38: 385-390.

Bryant, A.F. (1917) Influence of Heredity in Stammering. *Journal of Heredity* 8: 46-47.

Gutzman, N.H. (1924) Das Stottern. In *Sprachheilkunde*. (3rd Ed) Berlin. Fishers, Kornfield.

Orton, S.T. (1928) A Physiological Theory of Reading Disability and Stuttering in Children. *NEJM* 199: 1046-1052

West, R. (1929) *J Neurol and Psychopathology* 38:114.

Travis, L.E. (1931) *Speech Pathology*. New York. Appleton.

Wepman, J. (1935) Is Stuttering inherited ? *Amer Speech Correction Assoc.* Madison.

Wepman, J. (1939) Familial Incidence of Stammering. 30: 199-204

Nelson, S.E. (1939) The Role of Heredity in Stuttering. *Journal of Pediatrics* 14:642-654.

West, R., Nelson, S., Berry, M. (1939) The Heredity of Stuttering. *Q J Speech* 25: 23-30.

Gray, M. (1940) The "X" family: a Clinical and Laboratory Study of a "Stuttering" Family. *J of Speech Disorders* 5:343-348.

Johnson, W., King, A. (1942) An Angle Board and Hand Usage Study of Stutterers and Non-Stutterers. *J. Exp Psychol* 31: 293-311.

Meyer, B. (1945) Psychosomatic Aspects of Stuttering. *J Nerv Ment Dis* 101:127-157.

Sheehan, J.G., Martyn, M.M. (1970) Stuttering and its Disappearance. *J Speech Hearing Res* 13: 279-289.

Wingate, M.E. (1970) Stuttering, 1970: Where do We Stand ? *Journal of Speech and Hearing Research* 13:5-8

Quader, S.E. (1977) Dysarthria: an Unusual Side Effect of TCA. *Br Med J* 9:97.

Kidd, K.K. (1977) A Genetic Perspective on Stuttering. *J Fluency Disord* 2:259-269.

Sheehan, J.G.,Costley, M.S. (1977) A Reexamination of the Role of Heredity in Stuttering. *Journal of Speech and Hearing Disorders* 52:47-59.

World Health Organization (1977) *International Classification of Diseases*. p. 202

Hall, J.W., Jerger, J. (1978) Central Auditory Function in Stutterers. *J Speech and Hearing Research* 21: 324-337.

Overstake, C.P. (1979) *Stuttering: a New Look at an Old Problem Based on Neurophysiologic Aspects*. Springfield. Charles Thomas, pp. 14-34

Howie, P.M. (1981) Concordance for Stuttering in Monozygotic and Dizygotic Twin Pairs. *Journal of Speech and Hearing Research* 24:317-321.

Kidd, K.K., Heimbuch, R.C., Records, M.A. (1981) Vertical Transmission of Susceptibility to Stuttering with Sex-modified Expression. *Proc. Nat Acad. Sci.* 78:606-610.

Gladstien, K.L., Seider, R.A., Kidd, K.K. (1981) Analysis of the Sibship Patterns of Stutterers. *J Speech and Hearing Research* 24:460-462.

Nurnberg, H.G., Greenwald, B. (1981) Stuttering: an Unusual Side Effect of Phenothiazines. *Am J Psych* 138:386-7.

Kidd, K.K. (1982) Genetic Models of Stuttering. *J Fluency Disord* 7: 309-321

Van Riper, C.G. (1982) *The Nature of Stuttering*. Second edition.. Englewood Cliffs, NJ. Prentice-Hall.

Andrews, G.,Craig, A.,Feyer, A-M., et al. (1983) Stuttering: A Review of Research Findings and Theories Circa 1982. *J Speech Hear Dis* 48:226-246.

Seider, R.A., Gladstien, K.L., Kidd, K.K. (1983) Recovery and Persistence of Stuttering among Relatives of Stutterers. *J Speech and Hearing Disorders* 48:394-402.

Cox, N.J., Kramer, P.L., Kidd, K.K. (1984) Segregation Analysis of Stuttering. *Genetic Epidemiology* 1:245- 253.

Cox, N.J.,Seider, R.A.,Kidd, K.K. (1984) Some Environmental Factors and Hypothesis for Stuttering in Families with Several Stutterers. *J Speech Hear Res* 27:543-548.

Kidd, K.K. (1984) Stuttering as a Genetic Disorder. In Curlee, R.F. and Perkins, W.H. (editors.) *Nature and Treatment of Stuttering*. San Diego. College-Hill Press.

Shames, G.H., Rubin, H. (1986) *Stuttering: Then and Now*. Columbus. Charles E. Merrill.

Andrews, G.,Neilson, M.,Curlee, R. (1988) Stuttering. *JAMA* 260:1445

Cox, N.J. (1988) Molecular Genetics: The Key to the Puzzle of Stuttering ? *ASHA* 30:36-40.

Hays, P., Field, L.L. (1989) Postulated Genetic Linkage Between Manic-Depression and Stuttering. *Journal of Affective Disorders* 16:37-40.

Rosenfield, D.B. (1989) Stuttering. *NEJM* 320: 1630.

Leung, A.K.C., Robson, W.L.M. (1990) Stuttering. *Clinical Pediatrics* 29:498-502.

Andrews, G., Morris-Yates, A., Howie, P., Martin, N.G. (1991) Genetic Factors in Stuttering Confirmed. *Arch Gen Psych* 48:1034-1035.

Brady, J.P. (1991) The Pharmacology of Stuttering: A Critical Review. *Am J Psych* 148:1309-1316.

MacFarlane, W.B., Hanson, M., Walton, W., Mellon, C.D. (1992) Stuttering in Five Generations of a Single Family. *J. Fluency Disord* 16: 117-123.

The Genetics of Tourette Syndrome

Gilles de la Tourette (1885) Etude sur une affection nerveuse caractérisée par l'incordination motrice accompagnee d'écholalie et de coprolalié. *Arch Neur* (Paris) 9:158-200.

Fernando, S.J.M. (1967) Gilles de la Tourette Syndrome. *Brit J of Psych* 113:607-617.

Eldridge, R., Sweet, R., Lake, C.R., et al. (1977) Gilles de la Tourette Syndrome: Clinical, Genetic, Psychologic and Biochemical Aspects in 21 Selected Families. *Neurology* 27:115-124.

Golden, G.S. (1978) Tics and Tourettes's : A Continuum of Symptoms ? *Annals of Neurology* 4:145-148.

Kidd, K.K.,Prusoff, B.A.,Cohen, D.J. (1980) The Familial Pattern of Gilles de la Tourette Syndrome. *Arch Gen Psych* 37:1336-1339.

Nee, L.E.,Caine, E.D.,Polinsky, R.J., et al. (1980) Gilles de la Tourette syndrome: Clinical and Family Study of 50 Cases. *Ann Neurol* 7:41-49.

Pauls, D.L.,Cohen, D.J.,Heimbuch, R., et al. (1981) Familial Pattern and Ttransmission of Gilles de la Tourette Syndrome and Multiple Tics. *Arch Gen Psych* 38:1091-1093.

Baron, M., Shapiro, E., Shapiro, A., Ranier, J.D. (1981) Genetic Analysis of Tourette Syndrome Suggesting Major Gene Affect. *Am J Hum Gen* 33:767-775.

Montgomery, M.A., Clayton, P.J., Friedhoff, A.J. (1982) Psychiatric Illness in Tourette Syndrome Patients and First Degree Relatives. In Chase, T.N. & Friedhoff, A.J. (editors) *Gilles de la Tourette Syndrome*. New York. Raven Press.

Nee, L.E.,Polinsky, R.J.,Ebert, M.H. (1982) Tourette Syndrome: Clinical and Family studies. In Chase, T.N. & Friedhoff, A.J. (editors) *Gilles de la Tourette Syndrome*. New York. Raven.

Comings, D.E.,Comings, B.G., Devor, E.J., Cloninger, C.R. (1984) Detection of a Major Gene for Gilles de la Tourette Syndrome. *Amer J Hum Gen* 36: 586-600.

Comings, D.E.,Comings, B.G. (1984) Tourette Syndrome and Attention Deficit Disorder with Hyperactivity: Are they Genetically Related? J *Amer Acad of Child Psych* 23:138-146.

Pauls, D.L.,Kruger, S.D.,Leckman, J.F., et al. (1984) The Risk of Tourette's Syndrome and Chronic Multiple Tics Among Relatives of Tourette Syndrome Patients Obtained by Direct Interview. *Journal of the American Academy of Child Psych* 23:134-137.

Devor, E.J. (1984) Complex Segregation Analysis of Gilles de la Tourette syndrome: Further Evidence for a Major Locus Mode of Transmission. *Am J Hum Gen* 36:704-709.

Price, R.A.,Kidd, K.K.,Cohen, D.J., et al. (1985) A Twin Study of Gilles de la Tourette Syndrome. *Arch Gen Psych* 42: 815-820.

Comings, D.E.,Comings, B.G. (1985) Tourette Syndrome: Clinical and Psychological Aspects of 250 Cases. *Amer J Hum Gene* 37:435-450.

Pauls, D.L.,Leckman, J.F. (1986) The Inheritance of Gilles de la Tourette's Syndrome and Associated Behavior: Evidence for Autosomal Dominant Transmission. *NEJM* 315:993-997.

Pauls, D.L.,Hurst, C.R.,Kidd, K.K., et al. (1986) Tourette Syndrome and Attention Deficit Disorder: Evidence Against a Genetic Relationship. *Arch of General Psych* 43:1177-1179.

Pauls, D.L.,Towbin, K.E.,Leckman. J.F., et al. (1986) Gilles de la Tourette Syndrome and Obsessive Complusive Disorder: Evidence Supporting an Etiological Relationship. *Arch Gen Psych* 43:1180-1182.

Kurlan, R.,Behr, J.,Medved, L., et al. (1987) Severity of Tourette's Syndrome in One Large Kindred : Significance for Determination of Disease Prevalence Rate. *Arch of Neur* 44:268-269.

Cohen, D.J., Bruun, R.D., Leckman, J.F. (editors) (1988) *Tourette's Syndrome and Tic Disorders: Clinical Understanding and Treatment.* New York. Wiley.

Pauls, D.L., Leckman, J.F. (1988) The Genetics of Tourette's Syndrome. In Cohen, D.J., Bruun, R.D., Leckman, J.F. (editors) *Tourette's Syndrome and Related Tic Disorders*. New York. Wiley, pp. 92-101.

Comings, D.E. (1989) Presidential Address: The Genetics of Human Behavior-Lessons for Two Societies *Am. J. Hum. Gen.* 44:452-460.

Hasstedt, S.J., Leppert, M., et al. (1995) Intermediate Inheritance of Tourette Syndrome, Assuming Assortative Mating. *AJHG* 57:682-89.

Kanner, L. (1943) Autistic Disturbances of Affective Contact. *Nervous Child* 2:217-250.

Rutter, M. (1967) Psychotic Disorders in Early Childhood. *Br J Psych, Special Publication* No 1:113-158.

Hanson, D.R., Gottesman, I.I. (1976) The Genetics, If Any, of Infantile Autism and Childhood Schizophrenia. *J Autism Dev Disord* 6:209-234.

Folstein, S., Rutter, M. (1977) Infantile Autism. A Genetic Study of 21 Twin Pairs. *J Child Psychol Psych* 18: 297-321.

Deykin, E.Y., MacMahon, B. (1980) Pregnancy, Delivery,and Neonatal Complications Among Autistic Children. *AJDC* 134:860-64.

August, G.J., Stewart, M.A., Tsai, L. (1981) The Incidence of Cognitive Disabilities in the Siblings of Autistic Children. *Br J Psychiatry* 138:416-422.

Tsai, L.Y., Stewart, M.A., August, G. (1981) Implications of Sex Differences in the Familial Transmission of Infantile Autism. *J Autism Dev Disord* 11:165-173.

Minton, J., Campbell, M., Green, W.H., et al. (1982) Cognitive Assessment of Siblings of Autistic Children. *J Am Acad Child Adolesc Psych* 21:256-261.

Baird, T.D., August, G.J. (1985) Familial Heterogeneity in Infantile Autism. *J Autism Dev Disord* 15: 315-321.

Ritvo, E.R., Freeman, B.J., Mason-Brothers, A., et al. (1985) Concordance for the Syndrome of Autism in 40 Pairs of Affected Twins. *Am J Psych* 142: 74-77.

Ritvo, E.R., Spence, M.A., Freeman, B.J., et al. (1985) Evidence for Autosomal Recessive Inheritance in 46 Families with Multiple Incidences of Autism. *Am J Psych* 142:187-192.

Spence, M.A., Ritvo, E.R., Marazita, M.L., et al. (1985) Gene Mapping Studies with the Syndrome of Autism. *Behav Genet* 15:1-13.

Reiss, A.L., Feinstein, C., Rosenbaum, K.N. (1986) Autism and Genetic Disorders. *Schizophrenia Bull* 12:724-738.

Ritvo, E.R.,Mason-Brothers, A.,Menkes, J.H., et al. (1988) Association of Autism, Retinoblastoma and Reduced Esterase D Activity. *Arch Gen Psych* 45:600.

Smalley, S.L.,Asarnow, R.F.,Spence, M.A. (1988) Autism and Genetics. *Arch Gen Psych* 45:953-961.

The Genetics of Attention Deficit Hyperactivity Disorder

Cantwell, D. (1972) Psychiatric Illness in the Families of Hyperactive Children. *Arch Gen Psych* 27:414-417.

Morrison, J. (1980) Adult Psychiatric Disorders in Parents of Hyperactive Children. *Am J Psych* 137:825-27.

Biederman, J., Faraone, S., et al. (1990) Family - Genetic and Psychosocial Risk Factors in DSM -III Attention Deficit Disorder. *J Am Acad Child Adolesc Psych* 29: 526-533.

Stevenson, J. (1992) Evidence for a Genetic Etiology in Hyperactivity in Children. *Behav Genet* 22: 337-344.

Cook, E.H., Stein, M.A., et al. (1995) Association of Attention-Deficit Disorder and the Dopamine Transporter Gene. *AJHG* 56: 993-998.

The Genetics of Personality Disorders

Brown, F.W. (1942) Heredity in the Psychoneuroses. *Proc. Roy. Soc. Med.* 35:785-790.

Slater, E., Slater, P. (1944) A Heuristic Theory of Neurosis. *J. Ment. Psychiatr.* 7:49-55.

Eysenck, H.J., Prell, D.B. (1951) The Inheritance of Neuroticism: An Experimental Study. *J Ment Sci* 97:441.

Gottesman, I.I. (1963) Heritability of Personality: a Demonstration. *Psych. Monographs.* 77:1-21.

Shields, J., Slater, E. (1971) "Diagnostic Similarity in Twins with Neurosis and Personality Disorders." In Shields, J., Gottesman, I.I., (editors) *Man, Mind, and Heredity.* Baltimore. Johns Hopkins University Press, pp. 252-257.

Slater, E., Cowie, V. (1971) "Deviations of Personality and Neurotic Reactions." In *The Genetics of Mental Disorders.* London. Oxford, pp. 92-121.

Inouye, E. (1972) Genetic Aspects of Neurosis: a Review. *Int. J. of Ment. Health* 1:176-189.

Miner, G.D. (1973) The Evidence for Genetic Components in the Neuroses. *Arch. Gen. Psych.* 29:111-118.

Ranier, J. (1976) "The Genetics of Neurosis and Personality Disorder." In Sperber, M.A. and Jarvik, L.F. (editors) *Psychiatry and Genetics.* New York. Basic Books.

Holden, C. (1987) The Genetics of Personality. *Science* 237: 598-601.

CHAPTER SEVEN: STRATEGIES TO LOCALIZE THE GENETIC FACTORS IN BROAD BEHAVIORAL PHENOTYPES

The Role of Association Studies

Falk, C.T., Rubenstein, P. (1987) Haplotype Relative Risks: An Easy Reliable Way to Construct a Proper Control Sample for Risk Calculations. *Ann Hum Gen.* 51: 227-233.

Terwilliger, J.D., Ott, J. (1992) A Haplotype Based "Haplotype Relative Risk" Approach to Detecting Allelic Associations. *Hum Hered* 42: 337-346.

The Role of Linkage Studies in Localizing Genes

The Revolution in Gene Mapping Using Polymorphic DNA markers

Kan, Y.W., Dozy, A.M. (1978) Polymorphism of DNA Sequence Adjacent to Human ß-globin Structural Gene: Relationship to Sickle Mutation. *Proc Nat Acad Sci* 75:5631-35.

Wyman, A.R., White, R.L. (1980) A Highly Polymorphic Locus in Human DNA. *PNAS* 77: 6754-6758.

Botstein, D., White, R.L., Skolnick, M., Davis, R.W. (1980) Construction of a Genetic Linkage Map in Man Using Restriction Fragment Length Polymorphisms. *Am J Hum Genet* 32: 314-331.

Phillips, J.A., Parks, J.S.,Hjelle, B.L., et al. (1982) Genetic Analysis of Familial Isolated Growth Hormone Deficiency Type I. *J Clin Invest* 70:489-95.

Orkin, S.H., Kazazian, H.H., Antonarkis, S.E., et al. (1982) Linkage of ß-thalasemia Mutations and ß-globin Gene Polymorphisms with DNA Polymorphisms in Human ß-globin Gene Cluster. *Nature* 296:627-32.

Woo, S.L., Lidsky, A.S., Güttler, F., et al. (1983) Cloned Human Pheylalanine Hydroxylase Gene Allows Prenatal Diagnosis and Carrier Detection of Classical Phenoketonuria. *Nature* 306:151-54.

Camarino, G., Mattei, M.G., Mattei, J.F., et al. (1983) Close Linkage of Fragile X Mental Retardation Syndrome to Haemophilia B and Transmission Through a Normal Male. *Nature* 306: 701-704.

Tsipouras, P., Myers, J.C., Ramirez, F., et al. (1983) Restriction Fragment Length Polymorphism Associated with the Pro a2(I) Gene of Human Type

1 Procollagen. *J Clin Invest* 72:1262-67.

Davies, K.E., Pearson, P.L., Harper, P.S., et al. (1983) Linkage Analysis of Two Cloned DNA Sequences Flanking the Duchene Muscular Dystrophy Locus on the Short Arm of the Human Chromosome X. *Nucleic Acid Research* 11:2303-12.

Nussbaum, R.L., Crowder, W.E., Nyhan, W.L., et al. (1983) A Three -allele Restriction Fragment Length Polymorphism at the Hypoxanthine Phosphoribosyltransferace Locus in Man. *Proc Nat Acad Sci* 80:4035-39.

Jeffreys, A.J.,Wilson, V., Thein, S.L. (1985) Hypervariable 'Minisatellite ' Regions in Human DNA. *Nature* 314: 67-73.

White, R. (1985) DNA Sequence Polymorphisms Revitalize Linkage Approaches in Human Genetics. *Trends in Genetics.* 1:177-181.

Baehner, R.L., Kunkel, L.M., Monaco, A.P., et al. (1986) DNA Linkage Analysis of X chromosome- Linked Chronic Granulomatous Disease. *Proc Natl Acad Sci* 83:3398-3401.

Friend, S.H., Bernards, R., Rogeli, S., et al. (1986) A Human DNA Segment with Properties of the Gene that Predisposes of the Gene that Predisposes to Retinoblastoma and Osteosarcoma. *Nature* 1986; 323:643-646.

Monaco, A.P., Neve, R.L., Coletti-Feener, C, et al. (1986) Isolation of Candidate cDNA's for Portions of the Duchene Muscular Dystrophy Gene. *Nature* 323 646-650.

White, R.L., Lalouel, J-M., Lathrop, et al. (1987) Mapping Approaches to Gene Identification in Humans. *West J Med* 147:423-27.

Nakamura, Y., Leppert, M., O'Connell, P.,et al. (1987) Variable Number Tandem Repeat (VNTR) Markers for Human Gene Mapping. *Science* 235:1616-1621.

Linkage Basics

Sturtevant, A.H. (1913) The Linear Arrangement of Six Sex-Linked Factors in Drosophilia, as Shown by Their Mode of Association. *J Exp Zool* 14:43-59.

Suarez, B.K., Rice, J., Reich, T. (1978) The Generalized Sib Pair IBD Distribution : Its Usefulness in Detection of Linkage. *Annals of Human Genetics* 42:87.

Elston, R.C. (1984) Genetic Analysis Workshop II: Sib Pair Screening Tests for Linkage. *Genet Epid* 1: 175-178, 1984.

Blackwelder, W.C., Elston, R.C. (1985) A Comparison of Sib -pair linkage Tests for Disease Susceptibility Loci. *Genet Epid* 2:1985-97.

Weeks, D.E., Lange, K. (1988) The Affected-Pedigree-Member Method of Linkage Analysis. *Am J Hum Genet* 42:315-326.

Experimental Procedure

Lange, K., Boehnke, M. (1982) How Many Polymorphic Genes Will it take to Span the Human Genome ? *Am J Hum Genet* 34:842-45.
Renwick, J.H. (1971) The Mapping of Human Chromosomes. *Annual Review of Genetics*. Volume 5,p. 88.

Statistical Analysis of Linkage Data

Morton, N.E. (1955) Sequential Tests for the Detection of Linkage. *Am J Hum Genet* 7:277-318.
Haseman, J.K., Elston, R.C. (1972) The Investigation of Linkage Between a Quantitative Trait and a Marker Locus. *Behav Genet* 2: 3-19.
Ott, J. (1974) Estimation of the Recombination Fraction in Human Pedigrees: Efficient Computation of the Likelihood for Human Linkage Studies. *Am J Hum Genet* 26:588-597.
Kidd, K.K., Ott, J. (1984) Power and Sample Size in in Linkage Studies. Human Gene Mapping 7. *Cytogenet Cell Genet* 37:510-511.
Lathrop, G.M., Lalouel, J-M., Julier, J.M., Ott, J. (1985) Multilocus Linkage Analysis in Humans: Detection of Linkage and Estimation of Recombination. *Am J Hum Genet* 37:482-498.
Lathrop, G.M., Lalouel, J-M.(1988) Efficient Computations in Multilocus Linkage Analysis. *Am J Hum Genet* 42:498-505.
Paterson, A.H., Lander, E.S., et al. (1988) Resolution of Quantitative Traits into Mendelian factors by Using a Complete Linkage Map of Restriction Fragment Length Polymorphisms. *Nature* 335:721-26.

The Future of Linkage Studies in Psychiatric Genetics

Goldin, L.R., Gershon, E.S. (1983) Association and Linkage Studies of Genetic Marker Loci in Major Psychiatric Disorders. *Psychiat Dev* 4:387-418.
Kidd, K.K., Gerhard, D.S., Kidd, J.R., Houseman, D., Egeland, J.A. (1984) Recombinant DNA Methods in Genetic Studies of Affective Disorders. *Proc 14th C.I.N.P. Cong Clin Neuropharm* 7 (suppl 1) 198-199.
Gurling, H. (1985) Application of Molecular Biology to Mental Illness.: Analysis of Genomic DNA and Brain RNA. *Psych Dev* 3:257-273.

Stuart, E, Mcguffin, P. (1985) Can Linkage and Marker Association Resolve the Genetic Etiology of Psychiatric Disorders? Review and Argument. *Psychol Med* 15:455-462.

Cox, N.J., Suarez, B.K. (1985) Linkage Analysis for Psychiatric Disorders II. Methodologic Considerations. *Psych Devel* 4: 369-382.

Suarez, B., Cox, N. (1985) Linkage Analysis for Psychiatric Disorders: I - Basic Concepts. *Psychiatric Dev* 3:219-243.

Weissman, M.M., Merikanges, K.R., Wickramaratne, P., et al. (1986) Family-Genetic Studies of Psychiatric Disorders: Developing Technologies. *Arch Gen Psych* 43:1104-1116.

Gershon, E.S., Goldin, L.R. (1987) The Outlook for Linkage Research in Psychiatric Disorders. *J Psych Res* 21:541-550.

Kelly, T.E. (1987) Genetic Linkage : An Introduction to Basic Concepts. *Growth Genetics and Hormones* 3:1-10.

Kidd, K.K. (1987) Research Design Considerations for Linkage Studies of Affective Disorders Using Recombinant DNA Markers. *J Psychiat Res* 21: 551-57.

Kidd, K.K. (1987) *Searching for Major Genes for Psychiatric Disorders. Molecular Approaches to Human Polygenic Disease*. Ciba Foundation Symposium. Chichester. Wiley, p. 184-196.

Baron, M., Ranier, J.D. (1988) Molecular Genetics and Human Disease: Implications for Modern Psychiatric Research and Practice. *British Journal of Psych* 152:741-753.

Hyman, S.E. (1988) The Role of Molecular Biology in Psychiatry. *Psychosomatics* 29:328-332.

White, R, Lalouel, J.M. (1988) Investigation of Genetic Linkage in Human Families. *Advances in Human Genetics* Vol 16, pp. 121-228.

CHAPTER EIGHT: THE CURRENT STATUS OF ATTEMPTS TO LOCALIZE GENES FOR THE BROAD BEHAVIORAL PHENOTYPES

The Search to Localize a Gene in Schizophrenia

Association Studies in Schizophrenia

Irvine, D.,Miyashita, H. (1965) Blood types in Relation to Depressions and Schizophrenia. *Canadian Medical Assoc* J 92:551.

Rudduck, C.,Beckman, L., et al. (1985) C3 andC6 Complement Types in

Schizophrenia. *Hum Hered* 35: 255-258.
McGuffin, P., Sturt, E. (1986) Genetic Markers in Schizophrenia. *Hum Hered* 36:65-88.

Unique Clinical Cases

Constantinidis, J.K. (1958) Les marques de chromome chez les schizophrenes et la recherche du entre ces caracteres et la schizophrenie par la methode de Penrose. *J Genet Hum* 7:189.
Baron, M. (1976) Albinism and Schizophreniform Psychosis: A Pedigree Study. *Am J Psych* 133: 1070-73.
Book, J.A., Wetterberg, L., Modrzweska, K. (1978) Schizophrenia in a North Swedish Geographical Isolate. 1900-1977. Epidemiology, Genetics, and Biochemistry. *Clin Genet* 14:373-394.

Candidate Genes

Craig, S.P., Buckle, V.J., Craig, J.W., et al. (1986) Localization of the Human Tyrosine Hydroxylase Gene to Chromosome 11p15. Human Gene Mapping. *Cytogenet Cell Genet* 40:610.
Gurling, H. (1986) Candidate Genes and Favoured Loci: Stratagies for Molecular Genetic Research into Schizophrenia, Manic Depression, Autism, Alcoholism and Alzheimer's disease. *Psych Dev* 4:289-309.

Cytogenetic Abnormalities

Chodirker, B.N., Chudley, A.E., Ray, M., et al. (1987) Fragile 19p13 in a Family with Mental Illness. *Clinical Genetics* 31:1-6.
Bassett, A.S., McGillivray, B.C., Jones, B.D., Pantzar, J.T. (1988) Partial Trisomy Chromosome 5 Cosegregating with Schizophrenia. *Lancet* April 9, 1988 p. 799-801.

Linkage Studies in Schizophrenia

Elston, R.C., Kringlen, E., Namboodiri, K.K. (1973) Possible Linkage Relationships Between Certain Blood Groups and Schizophrenia and other Psychoses. *Behav Genet* 3:101-106.
Turner, W.J. (1979) Genetic Markers for Schizotaxia. *Biol Psych* 14:177-206.
McGuffin, P., Festenstein, H., Murray, R. (1983) A Family Study of HLA

Antigens and other Genetic Markers in Schizophrenia. *Psychol Medicine* 13:31-43.

Chadda, R., Kulhara, P., Singh, T., Sehgal, S. (1986) HLA Antigens in Schizophrenia : A Family Study. *Brit J Psych* 149:612-15.

McGuffin, P., Sturt, E. (1986) Genetic Markers in Schizophrenia. *Hum Hered* 36:65-88.

Andrew, B., Watt, D.C., Gillespie, C., Chapel, H. (1987) A Study of Genetic Linkage in Schizophrenia. *Psychological Medicine* 17:363-370.

Goldin, L.R., DeLisi, L.E., Gershon, E.S. (1987) The Relationship of HLA to Schizophrenia in 10 Nuclear Families. *Psych Res* 20: 69-77.

Byerley, W.F., LaLouel, J.M., et al. Utah Molecular Genetic Study of Schizophrenia. *APA New Research Abstracts*, 1988.

Gurling, H.M.D., Sherrington, R., Brynjolsson, J., et al. (1988) Genetic Linkage Study of Schizophrenia Using Minisatellite and Chromosome 5 Polymorphisms. *Schizophrenia Research* Vol 1:130.

Kennedy, J.L., Giuffra, L.A., Moises, H.W., et al. (1988) Evidence Against Linkage of Schizophrenia to Markers on Chomosome Five in a Northern Swedish Pedigree. *Nature* 336:167-170.

DeLisi, L.E., Detera-Wadleigh, S., et al. (1988) Genetic Marker Studies in Schizophrenia. *Schizophrenia Research* 1:128.

Sherrington, R., Brynjolfsson, J., et al. (1988) Localization of a Susceptibility Locus for Schizophrenia on Chromosome 5. *Nature* 336:164-167.

St Clair, D.,Blackwood, D.,Muir, W. (1988) Harvey-Ras-1 and Insulin Gene in Schizophrenia. Initial Findings in Five Scottish Pedigrees. *Schizophrenia Research* 1:131.

Lander, E.S. (1988) Splitting Schizophrenia. *Nature* 336:105-106.

St Clair, D. ,Blackwood, D., et al. (1989) No Linkage of Chromosome 5q11-q13 Markers to Schizophrenia in Scottish Families. *Nature* 339:305-309.

Coon, H., Jensen, S., et al. (1994) Genomic Scan for Genes Predisposing to Schizophrenia. *AJMG* 54:59-71.

Pulver, A.E., Karayiorgou, M., et al. (1994) A Sequential Strategy to Identify a Susceptibility Gene for Schizophrenia. I. Report of Potential Linkage on Chromosome 22q12-q13.1. *AJMG* 54:36-43.

The Current Status of Localizing Genes in Bipolar Disorder

Association Studies in Bipolar Disorder

Thomas, J.C., Hewitt, E.J.C. (1939) Blood groups in Health and Mental Disease. *J Ment Sci* 85:667-68.

Parker, J., Theille, A.,Spielberger, C. (1961) Frequency of Blood Types in a Homogenous Group of Manic Depressive Patients. *J Ment Sci* 107: 936.

Irvine, D.,Miyashita, H. (1965) Blood Types in Relation to Depressions and Schizophrenia. *Canadian Medical Assoc J* 92:551.

Tanna, V.L., Winokur, G. (1968) A Study of Association and Linkage of ABO Blood Types and Primary Affective Disorder. *Brit J Psychiat* 114: 1175-81.

Lim, L.C.C., Powell, J.F., et al. (1994) Monoamine Oxidase A Gene and Bipolar Affective Disorder. *AJHG* 54: 1122-1124.

Kawada, Y., Hattori, et al. (1995) Possible Association Between Monoamine Oxidase A Gene and Bipolar Affective Disorder. *AJHG* 56:335-336.

Lim, L.C.C., Powell, J.F., et al. (1995) Evidence of Genetic Association Between Alleles of MAOA Gene and Bipolar Affective Disorder. *AJMG* 60:325-331.

Nöthen, M.M., Eggermann, K., et al. (1995) Association Analysis of the Monoamine Oxidase A Gene in Bipolar Affective Disorder by Using Family-Based Internal Control. *AJHG* 57: 975-977.

Linkage Studies in Bipolar Disorder

Tanna, V.L., Winokur, G. (1968) A Study of Association and Linkage of ABO Blood Types and Primary Affective Disorder. *Brit J Psychiat* 114: 1175-81.

Winokur, G., Tanna, V.L. (1969) Possible Role of X-linked Dominant Factor in Manic-Depressive Disease. *Dis Nerv Syst* 30:89-94.

Reich, T., Clayton, P., Winokur, G. (1969) Family History Studies:V. The Genetics of Mania. *Amer J Psych* 125:1358.

Winokur, G., Clayton, P.J., Reich, T. (1969) Manic Depressive Illness. C.V. Mosby. St. Louis.

Mendlewicz, J., Fleiss, J.L., Fieve, R.R. (1972) Evidence for X-linkage in the Transmission of Manic-Depressive Illness. *JAMA* 222: 1624-1627.

Fieve, R.R., Mendlewicz, J., Fleiss, J.L. (1973) Manic Depressive Illness: Linkage with the Xg Blood Group. *Amer J Psych* 130:1355-1359.

Mendlewicz, J., Fleiss, J.L.(1974) Linkage Studies with X-Chromosome Markers in Bipolar (Manic-depressive) and Unipolar (depressive) Illnesses. *Biol Psych* 9:261-294.

Mendlewicz, J., Fleiss, J.L., Fieve, R.R. (1975) Linkage Studies in Affective Disorders: The Xg Blood Group and Manic Depressive Illness. In Fieve,

R.R., Rosenthal, D.,Brill, H. (editors) *Genetic Research in Psychiatry.* Baltimore. Johns Hopkins Press, 219-232.

Fieve, R.R., Mendlewicz, J., Ranier, J.D., Fleiss, J.L. (1975) A Dominant X-Linked Factor in Manic-Depressive Illness: Studies with Color Blindness. In Fieve, R.R., Rosenthal, D.,Brill, H. (editors) *Genetic Research in Psychiatry.* Baltimore. Johns Hopkins Press, 241-254.

Belmaker, R.H., Wyatt, R.J. (1976) Possible X-linkage in a Family with Varied Psychosis. *Isr Ann Psychiatry* 14: 345-53.

Gershon, E.S., Bunney, W.E. Jr (1976) The Question of X-Linkage in Bipolar Manic-depressive Illness. *J Psych Res* 13:99-117.

Baron, M. (1977) Linkage Between an X-chromosome Marker (deutan color blindness) and Bipolar Affective Illness. *Arch Gen Psych* 34: 721-25.

Johnson, G.F.S., Leeman, M.M. (1977) Analysis of Familial Factors in Bipolar Affective Illness. *Arch General Psychiatry* 24:1074-1083.

Smeraldi, E., Negri, et al. (1978) HLA System and Affective Disorders: A Sibship Genetic Study. *Tissue Antigens* 12:270.

Targum, S.D., Gershon, E.S., Van Eerdewegh, M., Rogentine, N. (1979) Human Leukocyte Antigen System not Closely Linked to or Associated with Bipolar Manic-depressive Illness. *Biological Psychiatry* 14:615.

Gershon, E.S., Targum, S.D., Matthysse, S., et al. (1979) Color Blindness not Closely Linked to Bipolar Illness: Report of a New Pedigree Series. *Arch Gen Psych* 36:1423-1430.

Leckman, J.F., Gershon, E.S., McGinniss, et al. (1979) New Data do not Suggest Linkage Between the Xg Blood Group and Bipolar Illness. *Archives of General Psychiatry* 36:1435.

Mendlewicz, J., Linowski, P., Guroff, J.J., et al. (1979) Color Blindness Linkage to Bipolar Manic Depressive Illness: New Evidence. *Arch Gen Psych* 36:1442-1447.

Gershon, E.S., Mendlewicz, J., et al (1980) A Collaborative Study of Genetic Linkage of Bipolar Manic-Depressive Illness and Red Green Color Blindness. *Acta Psych Scand* 61:319-38.

Mendlewicz, J., Linowski, P., Wilmotte, J. (1980) Linkage Between Glucose -6-phosphate -dehydrogenase Deficiency and Manic -Depressive Psychosis. *Br J Psych* 137:337-342.

Gershon, E.S. (1980) Nonreplication of Linkage to X-chromosome Markers in Bipolar Illness. *Arch Gen Psych* 37:1200.

Smeraldi, E., Bellodi, L. (1981) Possible Linkage Between Primary Affective Disorder Susceptibility Locus and HLA Haplotypes. *American Journal of Psychiatry* 138:1232.

Weitkamp, L.R., Stancer, H.C., et al. (1981) Depressive Disorders and HLA: A Gene on Chromosome 6 that can Affect Behavior. *New England Journal of Medicine* 305:1301,1981.

Baron, M., Ranier, J.D., Risch, N. (1981) X-linkage in Bipolar Affective Illness: Perspectives on Genetic Heterogeneity, Pedigree Analysis and the X-Chromosome Map. *J Aff Disorders* 3:141-157.

Kruger, S.D., Turner, W.J., Kidd, K.K. (1982) The Effects of Requisite Assumptions on Linkage Analyses of Manic-Depressive Illness with HLA. *Biol Psych* 17:1081-1099.

Risch, N., Baron, M. (1982) X-Linkage and Genetic Heterogeneity in Bipolar-Related Major Affective Illness: Reanalysis of Linkage Data. *Ann Hum Genet* 46:153-166.

Goldin, L.R., Clerget-Darpoux, F., Gershon, E.S. (1982) Relationship of HLA to Major Affective Disorder not Supported. *Psychiatry Research* 7:29-45.

Suarez, B.K., Crougan, J. (1982) Is the Major Histocompatibility Complex Linked to Genes that Increass Susceptibility to Affective Disorder ? A Critical Appraisal. *Psychiatry Research* 7:19-27.

Baron, M., Risch, N. (1983) X-Linkage in Affective and Schizoaffective Disorders: Genetic and Diagnostic Implications. *Neuropsychobiol* 8:304-311.

Del Zompo, M., Bocchetta, A., Goldin, L.R., Corsini, G.U. (1984) Linkage Between X-chromosome Markers and Manic -Depressive Illness: Two Sardinian Pedigrees. *Acta Psych Scand* 70: 282-287.

Kidd, K.K., Egeland, J.A., Moulthan, L., et al. (1984) Amish Study IV: Genetic Linkage Study of Pedigrees of Bipolar Probands. *Am J Psych* 141:1042-1048.

Gerhard, D., Egeland, J.A., Pauls, et al. (1984) Is the gene for Affective Disorder Located on the Short Arm of Chromosome 11. Am J Hum Genet 36: 3S.

Egeland, J.A., Gerhard, D.S., Pauls, D.L., et al. (1987) Bipolar Affective Disorders Linked to DNA Markers on Chromosome 11. *Nature* 325: 783-787.

Detera-Wadleigh, S.D., Berrettini, W.H., Goldin, L.R., et al. (1987) Close Linkage of c-Harvey *-ras*-1 and the Insulin Gene to Affective Disorder is Ruled Out in three North American Pedigrees. *Nature* 325: 806-808.

Hodgkinson, S., Sherrington, R., Gurling, H., et al. (1987) Molecular Genetic Evidence for Heterogeneity in Manic Depression. *Nature* 325:805-06.

Baron, M., Risch, N., Hamburger, R., et al. (1987) Genetic Linkage Between X-Chromosome Markers and Bipolar Affective Illness. *Nature* 326: 289-292.

Mendlewicz, J., Simon, P., Sevy, S., et al. (1987) Polymorphic DNA Marker of X-Chromosome and Manic Depression. *The Lancet* May 30, 1230-32.

Kidd, J.R., Egeland, J.A., Pakstis, A.J., et al. (1987) Searching for a Major Genetic Locus for Affective Disorder in the Old Order Amish. *Journal of Psychiatric Research.* 21:577-580.

Egeland, J.A.(1988) A Genetic Study of Manic-Depressive Disorder Among the Old Order Amish of Pennsylvania. *Pharmacopsychiatry* 21:74-5.

McGuffin, P. (1988) Major Genes for Major Affective Disorder? *Brit J Psych* 153: 591-596.

Gill, M., McKeon, P., Humphries, P. (1988) Linkage Analysis of Manic Depression in an Irish Family Using H-ras 1 and INS DNA Markers. *Journal of Medical Genetics* 25:634-37.

Waters, B., Sengar, D., Marchenko, I., et al. (1988) A Linkage Study of Primary Affective Disorder. *Brit J Psych* 152: 560-562.

Kelsoe, J.R., Ginns, E.I., et al. (1989) Re-evaluation of the Linkage Relationship Between Chromosome 11p Loci and the Gene for Bipolar Affective Disorder in the Old Order Amish. *Nature* 342:238-243.

Baron, M., Freimer, N.F., et al. (1993) Diminished Support for Linkage Between Manic-Depressive Illness and X-Chromosome Markers in Three Israeli Pedigrees. *Nature Genetics* 3:49-55.

Berrettini, W., Ferraro, T.N., et al. (1994) Chromosome 18 DNA Markers and Manic-Depressive Illness: Evidence for a Susceptibility Gene. *PNAS* 91:5918-21.

McMahon, F.J., Stine, O.C., et al. (1995) Patterns of Maternal Transmission in Bipolar Affective Disorder. *AJHG* 56:1277-86.

Pauls, D.L., Ott, J., et al. (1995) Linkage Analysis of Chromosome 18 Markers Do Not Identify a Major Susceptibility Locus for Bipolar Affective Disorder in the Old Order Amish. *AJHG* 57:636-43.

Stine, O.C., Xu, J., et al. (1995) Evidence for Linkage of Bipolar Disorder to Chromosome 18 with a Parent-of-Origin Effect. *AJHG* 57:1384-94.

Other Attempts to Localize Psychgenes

Alzheimer Disease

Chartier-Harlin, M-C., Crawford, F., et al. (1991) Early-Onset Alzheimer's Disease Caused by Mutations at Codon 717 of the ß-amyloid Precursor Protein Gene. *Nature* 353:844-846.

Goate, A., Chartier-Harlin, M-C., et al. (1991) Segregation of a Missense

Mutation in the Amyloid Precursor Protein Gene with Familial Alzheimer Disease. *Nature* 349:704-706.
Murrell, J., Farlow, M, et al. (1991) A Mutation in the Amyloid Precursor Protein Associated with Hereditary Alzheimer's Disease. *Science* 254:97-99.
Mullan, M, Crawford F., et al. (1992) A Pathogenetic Mutation for Probable Alzheimer's Disease in the APP Gene at the N-terminus of ß-Amyloid. *Nature Genetics* 1:345-347.
Schellenberger, G.D., Bird, T.D., et al. (1992) Genetic Linkage Evidence for Familial Alzheimer's Disease Locus on Chromosome 14. *Science* 258:668-71.

Autism

Hallmayer, J., Pintado, E., et al. (1994) Molecular Analysis and Test of Linkage Between the FMR-1 Gene and Infantile Autism in Muliplex Families. *AJHG* 55:951-959.

Alcohol Dependence

Cloninger, C.R., Bohman, M., Sigvardsson, S. (1981) Inheritance of Alcohol Abuse. Cross Fostering of Adopted Men. *Arch Gen Psych* 8:861-868.
Goodwin, D.W. (1985) Alcoholism and Genetics. *Arch Gen Psych* 42:171-175.
Schuckit, M.A., Li, T.L., Cloninger, C.R., Deitrich, R.A. (1985) Genetics of Alcoholism. *Clin Exp Res* 9: 475-492.
Goldman, D. (1988) Molecular Genetic Markers for Linkage of Genetic Loci Contributing to Alcoholism. *Recent Dev Alcoh* 6:333-49.
Blum, K., Noble, E. (1990) Allelic Association of Human Dopamine D2 receptor gene in Alcoholism. *JAMA* 263: 2055-2060
Gelertner, J., O'Malley, S., et al. (1991) No Association Between the an Allele at the D2 Dopamine Receptor Gene (DRD2) and Alcoholism. *JAMA* 266:1801-1807.
Gelertner, J., Goldman, D., et al. (1993) The A1 Allele at the D2 Dopamine Receptor Gene and Alcoholism: a Reappraisal. *JAMA* 269:1673-1677.

Panic Disorder

Pauls, D.L., Bucher, K.D., Crowe, R.R., Noyes, R. (1980) A Genetic Study of Panic Disorder Pedigrees. *Am J Hum Genetics* 32: 639-644.
Crowe, R.R., Noyes, R., Pauls, D.L., Slyman, D. (1983) A Family Study of

Panic Disorder. *Arch Gen Psych* 40: 1065-69.
Surman, O.S., Sheehan, D.V., Fuller, T.C., Gallo, J. (1983) Panic Disorder in Genotypic HLA Identical Sibling Pairs. *Am J Psych* 140:237-239.
Torgersen, S. (1983) Genetic Factors in Anxiety Disorders. *Arch Gen Psych* 40: 1085-1089.
Noyes, R., Crowe, R.R., Harris, E.L., et al. (1986) Relationship Between Panic Disorder and Agoraphobia: A Family Study. *Arch Gen Psych* 43:227-232.
Crowe, R.R., Noyes, R. Jr., Wilson, A.F., et al. (1987) A Linkage Study of Panic Disorder. *Arch Gen Psych* 44: 933-37.
Kendler, K., Heath, A.C., Martin, N.G, Eaves, L.J. (1987) Symptoms of Anxiety and Symptoms of Depression : Same Genes Different Environment? *Arch Gen Psych* 44:451-57.Mutchler, K., Crowe, R.R.,
Noyes, R., et al. (1990) Exclusion of the Tyrosine Hydroxylase Gene in 14 Panic Disorder Pedigrees. *Am J Psych* 147:1367-9.

Stuttering

Mellon, C.D., Umar, F., Hanson, M.L. (1993) Stuttering as Phenotype for Behavioral Genetics Research. *AJMG* 48:179-183.

Dyslexia

Cardon, L.R., Smith, S.D., et al. (1994) Quantitative Trait Locus for Reading Disability on Chromosome 6. *Science* 266:276-9.

Speculations on the Failure to Localize Genes for the Broad Behavioral Phenotypes

Risch, N. (1990) Genetic Linkage and Complex Diseases, with Special Reference to Psychiatric Disorders. *Genet Epidem* 7:3-16.
Rose, S. (1995) The Rise of Neurogenetic Determinism. *Nature* 373:380-2. Craddock, N., Khodal, V., et al. (1995) Mathematical Limits of Multilocus Models: The Genetic Transmission of Bipolar Disorder. *AJHG* 57:690-702.

CHAPTER NINE:
SPECIAL CONCEPTS IN PSYCHIATRIC GENETICS

Falconer, D.S. (1967) *Introduction to Quantitative Genetics*. New York. Ronald Press. pp. 129-150.

Jacob, F. (1982) *The Possible and the Actual.* New York. Pantheon. p. 66.

CHAPTER TEN: IMPLICATIONS FOR THE PRACTICE OF PSYCHIATRY

Graham, J.M. (1988) *Smith's Recognizable Patterns of Human Deformation.* Philadelphia. Saunders. Second edition.

Bennett, R.L., Steinhaus, K.A., et al. (1995) Recommendations for Standardized Human Pedigree Nomenclature. *AJHG* 56:745-752.

Genetic Counseling

D'Arcy, E. (1968) Congenital Defects. Mother's Reaction to First Information. *Br. Med. J.* 3:796-798.

Hecht, F., Holmes L.B. (1972) What We Don't Know about Genetic Counseling. *NEJM* 286:464-465.

Reiss, J.A., Menasche, V. (1972) Genetic Counseling and Congenital Heart Dsease. *J. Pediatr.* 80:655.

Leonard, C., Chase, G.,Child, B. (1972) Genetic Counseling: A Consumers View. *NEJM* 287:433.

Antley, R.M., Hartage, L.C. (1976) Psychological Responses to Genetic Counseling for Down's Syndrome. *Clin. Genet.* 9:257-265.

Sperber, M.A. (1976) "Psychology and Metacommunication in Genetic Counseling." In Jarvik, L., Sperber, M.A. - *Psychiatry and Genetics.* New York. Basic Books.

Tsuang, M.T. (1978) Genetic Counseling for Psychatric Patients and their Families. *Am J Psych* 134: 1465-1475.

Kessler, S. (1979) *Genetic Counseling: Psychological Dimensions.* New York. Academic Press.

Targum, S.D.,Gershon, E.S. (1980) Genetic Counseling for Affective Illness. In Belmaker, R.H., Van Pragg, H.M. (editors) *Mania: an Evolving Concept.* New York. Spectrum Publications Inc.

Mikkelsen, M., Stene, E., Stene, J. et al. (1981) "The Patient with a Chromosomal Disorder, the Parents and the Professional Team." In Schmid, W. & Nielsen, J. (editors) *Human Behavior and Genetics.* Amsterdam. Elsevier, pp. 9-20.

Fresco, N., Silvestre, D. (1981) "Imaginary Constructions of Patients around the Words 'Genetics' and 'Heredity'." In Schmid, W. & Nielsen, J. (editors) *Human Behavior and Genetics.* Amsterdam. Elsevier, pp. 21-30.

Index

Acquired traits, 52
Acute intermittent porphyria, 70
Adoption studies, 92-94
Alcoholism, 123-124, 173-174
Alzheimer disease, 172-173
Amyotrophic lateral sclerosis, 81
Androgen insensitivity, 78
Anticipation, 50-51
Anxiety disorder, 32
Association studies, 141-143
Assortative mating, 52-53
Attention deficit hyperactivity disorder, 31, 135-136
Autism, 131-135, 173
Behavior
    abnormal behavior, 19-22
    constraints on, 17- 18
    definition, 15-16
    factors determining, 18-19
Behavioral phenotypes
    limited number, 34
Behavioral Teratogenesis, 189
Biological individuality, 49-50
Bipolar disorder, 25-27, 116-120, 160-172
Brief reactive psychosis, 192
Cat eye syndrome, 66
Cell cycle, 36-37
Chromosome
    addition of chromosome, 57-58
    deletion, 61-62
    duplication, 66
    loss of chromosome, 56

nomenclature, 55
number, 37
Chromosomal organization, 41
Coffin-Lowry syndrome, 74
Congenital deformation, 189
Cri-du-chat syndrome, 62-63
Dementia, 30-31
Dentatorubral-Pallidoluysian atrophy, 71
Developmental plasticity, 47, 186
Disinhibition-Dementia-Parkinsonism, 73
DNA, 38
Dominant inheritance, 42-43
Dyslexia, 174
Dysmorphology, 194-195
Edwards syndrome, 58
Familiality of very rare disorders, 48
Family history, 194
Family studies, 83-87
Fragile X, 76-77
Galactosemia, 80-81
Genetic counseling, 196-199
Genetics
basic concepts, 35
code, 38-40
Hallervorden-Spatz, 79
Hardy Weinberg equilibrium, 51-52
High risk studies, 94-95
Huntington disease, 69
Hurler syndrome, 80
Jacobsen syndrome, 63
Klinefelter syndrome, 59-60
Learning disabilities, 32-33, 136
Lesch-Hyan, 75
Life events, 190
Linkage studies, 144-153
Low recurrence risk disease, 46
Major depression, 27-29
Menkes disease, 75
Mental retardation, 29-30, 97-102

Metachromatic leukodystrophy, 73-74
Miller-Dieker syndrome, 64-65
Mucopolysaccharidosis, 77-78
Multifactorial threshold model, 47
Multiple personality disorder, 191
Mutation, 40-41
Myotonic muscular dystrophy, 73
Neurofibromatosis, 72-73
Niemann-Pick, 79-80
Oculocerebrorenal syndrome, 74
Olivopontocerebellar degeneration, 79
Ornithine Transcarbamylase, 75
Panic disorder, 174
Parent of origin effects, 53
Personality disorder, 33, 137-138
Phenocopies, 187-188
Phenylketonuria, 70-71
Polymorphism, 49
Post traumatic stress disorder, 191
Prader Willi, 63
Predisposition, 44
Primary degenerative dementia, 121-123
Recessive inheritance, 43-44
Rett syndrome, 80
Sanfillipo, 80
Schizophrenia, 22-25, 103-115, 155-160
Segregation analysis, 95
Service organizations, 201-202
Sex, 48
Simple versus complex disease, 45-46
Sjögren-Larsson syndrome, 73
Smith Lemli Opitz syndrome, 81
Smith Magenis syndrome, 65
Stuttering, 124-129, 174
Substance abuse, 33-34
Supermale phenotype, 60-61
Tay-Sachs disease, 72
Tourette syndrome, 129-
Triple X female, 61

Trisomy 8 mosaicism, 58
Trisomy 13, 58
Trisomy 21, 58-59
Tuberous Sclerosis, 78
Turner syndrome, 56-57
Twin studies, 87-91
Unipolar disorder, 120-121
Variation, 177
Velo-Cardio-Facial syndrome, 65
Williams syndrome, 63
Wilson disease, 71
Wolf-Hirschorn syndrome, 62
X - linked hydrocephalus, 77
X - linked inheritance, 44
X - linked mental retardation, 74

About the Author

Charles David Mellon, M.D. received undergraduate and masters degrees from the University of Texas at Austin and an M.D. from the University of Texas Health Science Center at San Antonio. Dr. Mellon is a board certified psychiatrist and is a member of the American Society of Human Genetics and the International Society of Psychiatric Genetics.

Notes

Notes

Notes